D1251925

COVID-19

CANADIAN ESSENTIALS
Series editor: Daniel Béland

Provocative thinking and accessible writing are more necessary than ever to illuminate Canadian society and to understand the opportunities and challenges that Canada faces. A joint venture between McGill-Queen's University Press and the McGill Institute for the Study of Canada, this series arms politically active readers with the understanding necessary for engaging in – and improving – public debate on the fundamental issues that have shaped our nation. Offering diverse and multidisciplinary perspectives on the leading subjects Canadians care about, Canadian Essentials seeks to make foundational and cutting-edge knowledge more accessible to informed citizens, practitioners, and students. Each title in this series aims to bolster individual action in order to support a better, more inclusive and dynamic country. Canadian Essentials welcomes proposals for concise and well-written books dealing with far-reaching and timely Canadian topics from a broad swath of authors, both within and outside of academia.

1 COVID-19
A History
Jacalyn Duffin

COVID-19

A History

JACALYN DUFFIN

McGill-Queen's University Press
Montreal & Kingston • London • Chicago

ISBN 978-0-2280-1411-9 (cloth)
ISBN 978-0-2280-1508-6 (ePDF)
ISBN 978-0-2280-1509-3 (ePUB)

Legal deposit fourth quarter 2022
Bibliothèque nationale du Québec

Printed in Canada on acid-free paper that is 100% ancient forest free
(100% post-consumer recycled), processed chlorine free

Funded by the   Financé par le
Government   gouvernement
of Canada   du Canada

Canada

Canada Council   Conseil des arts
for the Arts   du Canada

We acknowledge the support of the Canada Council for the Arts.
Nous remercions le Conseil des arts du Canada de son soutien.

Library and Archives Canada Cataloguing in Publication

Title: COVID-19 : a history / Jacalyn Duffin.
Names: Duffin, Jacalyn, author.
Description: Series statement: Canadian essentials; 1 | Includes bibliographical
references and index.
Identifiers: Canadiana (print) 20220272808 | Canadiana (ebook) 20220272891
| ISBN 9780228014119 (cloth) | ISBN 9780228015086 (ePDF) | ISBN 9780228015093
(ePUB)
Subjects: LCSH: COVID-19 (Disease)—History.
Classification: LCC RA644.C67 D84 2022 | DDC 614.5/92414—dc23

This book was typeset in Minion Pro.

For

essential workers everywhere

and

the memory of
those who could not be saved

Contents

Tables and Figures

Acknowledgments

First, many thanks go to Philip Cercone for suggesting that I write this book and believing that someone might want to read it. I hope he is right. Together with everyone else, I owe a huge debt of gratitude to the public-health workers, health-care professionals, and essential workers in grocery stores, garbage collection, food processing, and delivery, who cared for the sick and kept the rest of us safe and supplied. The dedicated professionals and fellow volunteers at Kingston, Frontenac, Lennox and Addington Public Health, led by Angela McKercher-Mortimer, always remained cheerful and patient, even when their daunting tasks became overwhelming; they taught me so much.

This book was written mostly under stay-at-home orders. Nevertheless, I visited Queen's University Library multiple times every day via the internet, relying on its collections, databases, and superb interlibrary loan, which provided ready access to valuable resources. Colleagues were quick to provide information and answer questions, in particular Henry Averns, Gerald Evans, Ian Gemmill, Kieran Moore, Kanji Nakatsu, Howard Shapiro, Wendy Wobeser, and Dick Zoutman.

Through Facetime, Zoom, email, phone, and Jamulus, friends and family were encouraging and inspiring. They delivered poignant stories, good humour, juicy commentary, useful tidbits, excellent edits, and lots of music that kept my spirits up and research coffers full. For these gifts and more, I thank Denise Bowes, Ross Duffin, Barbara Duffin-Bates, François Gallouin, Bert Hansen, Beate Huber, Anita Johnston, William Leiss, John Matthews, Joan Schwartz, Mindy Schwartz (not related), Beverly Simmons, Tom Stevinson, Bob Tennent and his recorder ensemble, the extended Wolves, Cherrilyn Yalin, Josh Lipton-Duffin, Jennifer Macleod, Jessica Wolfe, Daniel Goldbloom, and my grandchildren

Tycho, Maxwell, Wolf, and Kensie whom I miss so much. Two anonymous readers for the press offered many helpful suggestions with alacrity and tact, while eagle-eyed Louise Piper and Neil Erickson handled the production with patience and care.

Robert David Wolfe has been the maestro and mainstay of our two-person bubble. Without his gentle but firm guidance, his wisdom, vigilance, criticism, forbearance, and keen sense of humour, outrage, and wonder, this book would not exist.

COVID-19

COVID-19 Unmasked

Historians of medicine normally lead very quiet lives. Whenever a new disease appears, especially one that is contagious, we suddenly become the most sought-after pundits of the moment. I am old now, and I have seen this transformation happen over and over again with the advent of previously unknown, infectious conditions: toxic shock syndrome, Legionnaires' disease, Acquired immunodeficiency syndrome (AIDS), Ebola, Severe acute respiratory syndrome (SARS), H1N1, West Nile, and Zika. Even the 2018 centennial of Spanish flu provoked urgent appeals to historians from physicians, journalists, television anchors, radio hosts, and independent bloggers, seeking lessons from the past. The arrival of COVID-19 was no different.

New diseases serve as a reminder of prior scourges. There are many examples. In the fifth century BC, Thucydides wrote of the famous "Plague of Athens," and Hippocrates described the mysterious "Cough of Perinthus." Numerous writers have documented pandemic waves of bubonic plague from the seventh century through to the twentieth, focusing especially on fourteenth-century Europe. References to these past plagues happen so frequently that I once devised the "Thucydides index" as a way of identifying those years that had seen new diseases: the number of scientific citations to the ancient historian Thucydides increase in times of novel epidemics.

The biggest lessons from the past are simple: nature is endlessly inventive. New infectious agents will always arise, and our ability to confront them will be based on the best wisdom derived from our experiences with great scourges of the near and distant past. In thinking about how to deal with epidemics, we might even contemplate our own role in creating space for new diseases to evolve or spread.

Many books have been written about the power of infectious disease to affect the course of human history, from the famous *Plagues and Peoples* (1976) of William H. McNeill to Canadian Andrew Nikiforuk's *The Fourth Horseman* (1991) or his *Pandemonium* (2008). Exemplified by Charles E. Rosenberg's *The Cholera Years* (1962), histories of the mortality and social disruption owing to specific infectious diseases are legion. By the mid-1990s, books, such as *The Coming Plague* (1994) by Pulitzer-Prize-winner Laurie Garrett, were claiming that we were due if not overdue for another big pandemic and that the wondrous possibilities of air travel would enhance its spread. In 1999, well before SARS, the World Health Organization (WHO) began urging its member nations to strengthen their pandemic preparedness.

Back in 2000, the popular CBC Radio science program, *Quirks & Quarks,* was celebrating its 25th anniversary with a special show before a live audience in the Glenn Gould studio in downtown Toronto. The playful premise was that it was actually 2025, and the host, Bob McDonald (still the *Quirks* host today), had been cloned and brought back to be informed by the show's regular columnists about what had happened in their various branches of science during the fictitious previous quarter century. In other words, we were to predict future events. I was responsible for medicine. Out of my few minutes, which included fabricated comments on the human genome project (failed owing to a Supreme Court case), successful microsurgeries, and new cancer treatments, came the following three-minute conversation:

CBC Radio's *Quirks & Quarks,* 13 October 2000, at thirty-three minutes:
B: So, Jackie what was the biggest medical story that happened in the last twenty-five years?
J: Well, Bob, I hate to break this to you, but the biggest story was the Plague of 2013.
B: Plague of 2013! What happened?
J: There was a huge plague, seventy million people died. It swept the world in fact, but North America suffered most.
B: So, what caused it?
J: It was caused by a bacterium – an ordinary bacterium. You might remember that back in the year 2000 there were some multidrug resistant bacteria.

B: Superbugs.

J: Yes. And we knew that they had come along, encouraged by the use of pharmaceuticals, antibiotics. It was a big story in the middle of the 1990s back in the last century. Well, these superbugs were capable of creating horrible diseases in people, though mostly they were kept under control at a low level. But there was a massive eco-logical change when we developed the cure for the common cold.

B: Oh! We got the cure for the cold?

J: Yes. In the year 2012. It was widely used, it wiped out the cold, and it seems that, in the pathocenosis of the world at that time, this superbug moved in to take advantage. In the world-wide statistics you can see that North America suffered proportionately more than did other places in Asia and Africa.

B: How did solving the common cold lead to the plague?

J: It created space in the pathogens that affect us.

B: Oh! They moved in to where the common cold would have been.

J: Right. But the immediate cause of it was the multidrug resistant bacterium.

B: So how did the epidemic end?

J: Many people died. There was a lot of chaos, because the resources to support these people in the hospitals and intensive-care units were completely overrun. Caregivers caught the plague as well. People were dying in the streets. But within a year it had basically vanished. Those who had had the disease and lived were immune, those who didn't catch it were lucky, and the ones who died, were dead. The pharmaceu-tical industry was able to identify a new antibiotic for that particular strain.

B: Oh! So they've got a new antibiotic for the new disease that came along because of antibiotic resistance? Does that mean that it can all happen again?

J: It could all happen again.

B: That's spooky! Now I'm depressed.

Okay, I got it wrong – blaming bacteria instead of a virus and predicting a cure for the cold. On the other hand, much of it rings true: the over-whelmed hospitals, the great burden in North America, and the cure, though not a vaccine, came from industry. I was pleased to invoke the term pathocenosis, coined by my late mentor, Mirko Drazen Grmek, to indicate the ensemble of diseases that prevail in a certain time at a certain place. I wish I'd added climate change to the mix. Mirko Grmek had already been worried about that too back in the 1980s.

This radio conversation took place several years before SARS, and fully two decades before a man with unusual pneumonia entered a Wuhan hospital. At the after-show reception in the CBC lobby, I was mobbed by people worried about the possibility of such a plague and perplexed that we, safe in Canada, might be more vulnerable than others elsewhere – partly because of our privilege and freedoms. In the days following, I heard from similarly concerned listeners across the country wanting guidance on how to avoid the future, fictitious disease. The *Quirks* producers told me that I had been the only guest to predict a dark time ahead.

But this prediction was not especially prescient. Any other historian of medicine would have said something similar. I was struck by how much we all want to believe in our invulnerability, and how swiftly and completely, denial can enhance the danger.

Fast forward twenty years, to a time when I thought my final book was behind me. McGill-Queen's University Press editor Philip Cercone called to suggest that I write a history of COVID-19. I needed some time to think about his request. Mostly I like writing about things that are well and truly over. The pandemic would definitely not be over by 2022. Can one write a history of something that is happening right now? It seemed more like journalism, or hubris.

On the other hand, I was already watching the pandemic closely with the odd double vision of a physician who is also a historian. In the early 1980s, I had seen my elderly mentor, Grmek, a noted expert on diseases of antiquity, turn to writing a history of AIDS at its beginnings – a book that I had translated back in 1990. AIDS is not yet over. Thanks also to Grmek, I had become a historian of disease concepts, had spent thirty years teaching the history and philosophy of medical ideas, and had published on the histories of AIDS, hepatitis, and SARS. If, at the end of his career, Grmek could write about an entirely new disease caused by a novel pathogen, perhaps it was somehow fitting that I should do the same at the end of mine. And how strange that he and I, both, were confronted in early retirement with deadly catastrophes of global magnitude. With some trepidation, I signed a contract on 6 November 2020.

COVID-19 is definitely not yet over. But already it has a history and a wealth of sources. For a word that did not even exist until it was born on 11 February 2020, a search in Amazon for "COVID-19" produced more than twenty thousand books that had appeared in the nine months preceding the day that I signed that contract – an impressive gestation! By

February 2021, that number had risen to more than thirty thousand books (Amazon stops counting at thirty thousand). In October 2021, the same search in the Web of Science database produced more than 185,000 articles – all in the peer-reviewed, scientific literature.

My other sources had to be the news media from around the world providing early and ever-changing reports about the progress of the pandemic, its science, and its social and economic impact. These sources frequently contradict each other. Numerous historical accounts of COVID-19 already exist, explaining what happened and what went wrong, including Richard Horton's impassioned essay of May 2020, informed by his insights as editor of the *Lancet*, the journal that had published the earliest scientific articles. This massive outpouring of scholarship, journalism, and reporting constituted a daunting array of sources as a backbone to my project. A list of some sources used can be found at https://www.mqup.ca/filebin/pdf/Sources_to_COVID-19.pdf.

Covering the events of roughly two and a half years, the twelve chapters necessarily overlap in terms of chronology. The book is divided into three parts. The first part describes the pandemic from its origins to its extent around the globe. The different experiences of a few countries in various geographic regions are highlighted to serve as exemplars.

The second part pauses to explore the history and context of the medical science and discoveries that grew out of the COVID-19 experience in terms of cause, testing, controls, treatments, and vaccines. Here, more than elsewhere, I was channeling the methods of Mirko Grmek. He had wrestled with the many ways in which a disease could be (or could seem to be) new; and he coined the term "emerging disease." Grmek envisaged a mechanism by which disruptions to the balance of a pathocenosis, described above, can come from nature or from humans, making room for new diseases to arise. He taught that human achievements – sometimes wonderful achievements, such as controlling or eradicating diseases with drugs and vaccines – can have nefarious consequences further down the road. But the message is unwelcome, if not ignored.

The third part returns to the global progress of the pandemic in space and in time forward and in time past. It closes with some of the challenges that COVID-19 leaves for now and for our future.

Jacalyn Duffin
Kingston, Ontario
10 May 2022

PART ONE

The First Wave

"Behold, a pale horse! Its rider's name was Death, and Hades followed him. They were given power over a fourth of the earth to kill by sword, famine, and pestilence."

Revelation 6:8

1

It Begins

On 18 December 2019, Dr Ai Fen examined a man suffering from atypical pneumonia, with fever, cough, difficulty breathing, and a blotchy appearance on scan of his lungs. She was head of emergency at the Central Hospital in Wuhan, a city of eleven million in Hubei province, central China. Her patient worked at the Huanan (South China) Seafood Market. In the following days, doctors in Wuhan were puzzled by several other cases of unexplained pneumonia – unexplained, in that the sick had previously been healthy, tests for usual causes were negative, and antibiotics didn't help. Word of the outbreak began to circulate.

On 27 December, Dr Ai saw another patient with the same mysterious pneumonia but without any connection to the market. Virology test results from that patient came back on the afternoon of 30 December, listing a wide variety of possible causes – both viral and bacterial – including SARS coronavirus. She well remembered SARS (severe acute respiratory syndrome). That highly contagious coronavirus disease had swept around the world from China to Canada in 2003, infecting 8,100 and killing 10 per cent of its sufferers, before it suddenly, inexplicably stopped. Despite best efforts at the time, no vaccine or treatment had ever been found.

Dr Ai circled the words "SARS coronavirus," alerted her hospital's public-health department, snapped a photo, and sent it by text to another doctor in Wuhan. The news spread. Her young colleague, thirty-three-year-old ophthalmologist Dr Li Wenliang, saw the report with the highlighted words. He sent it through WeChat to a few medical school classmates, warning them to take precautions and wear safety gear to protect themselves and their families. Li and his wife, optometrist Fu Xuejie, had a five-year-old son and were expecting their second child. He did not usually look after people with breathing problems, but he had already heard about cases of strange pneumonia, and, as an ophthalmologist, he had to get very close to his patients. He asked his friends to share the information cautiously, but it was picked up by social media and disseminated widely.

Very late that same day, 30 December, the Chinese Centers for Disease Control sent an announcement to the global information sharing network, ProMED-mail, to report observations on four individuals, indicating that the first patient had worked in the Huanan Seafood Market, where live animals were sold. The source suggested animal-to-human spread, as is common for some viral diseases. An update on the following day reported twenty-seven patients in Wuhan. These communications are how the world became aware.

On 1 January, Dr Ai again contacted her hospital's public-health department, worried that the infection could affect her colleagues and hamper their ability to work. Soon after, she was severely reprimanded by officials at her hospital for spreading rumours. Nevertheless, also on 1 January, the Huanan market was closed, it was stated, for renovation and disinfection.

Two days later, on 3 January, Dr Li was summoned to the local office of the Public Security Bureau where he was accused of "making false comments" that had "severely disturbed the social order." He was forced to sign a letter admitting to "illegal activities" and promising not to spread more rumours.

Both Dr Ai and Dr Li kept working. On 11 January, Dr Ai learned that a nurse in her emergency department had contracted the pneumonia. She called a staff meeting to announce the strong possibility of human-to-human spread. That same day, China reported its first death from the new disease. Five days later, hospital officials were still not convinced that it could spread from one human to another.

For his part, on 8 January, Dr Li examined a healthy man with glaucoma who worked at the Huanan Seafood Market. The man developed a fever the following day, which Dr Li suspected might be the new disease. Sure enough, by 10 January, Dr Li spiked a temperature and began coughing. His condition deteriorated rapidly, and two days later, he was admitted, isolated, and treated in the intensive-care unit (ICU) of his own hospital. He was formally diagnosed with the new coronavirus infection three weeks later, on 1 February. Despite the official pressure, Dr Li kept sending messages and selfies from the hospital bed that became his deathbed sometime during the night of 6–7 February 2020.

By that time, millions of Chinese people were following his story, cheering for him, and then mourning him as a hero for free speech.

With the news of his death, they darkened their homes, waved glitter from their balconies, and blew whistles in his honour. Dr Li's grieving parents spoke of their wonderful son and demanded an explanation for his harsh treatment by police. An official apology was issued much later in March 2020. In early June, Dr Li's wife gave birth to a second son. She shared a picture of her baby on social media with the words, "Husband, can you see this from heaven? You have given me your final gift today. I will of course love and protect them."

The delay in Dr Li's formal diagnosis was not owing to any further suppression, but to the fact that when he fell ill, the virus was still under investigation. On 5 January, scientists in Shanghai had established that the cause was a novel (new) coronavirus, related to, but different from those that had caused SARS in 2003 and MERS (Middle East Respiratory Syndrome) in 2012. On the same day, the head of that laboratory fifty-five-year-old Zhang Yongzhen of Fudan University in Shanghai uploaded the viral genetic sequence to the US National Center for Biotechnology Information (NCBI). He alerted the Ministry of Health and travelled 850 kilometres to Wuhan to learn more of the clinical appearances. But Zhang knew that it could sometimes take weeks for other experts to notice new information at the NCBI site. After discussing the situation with virologist Edward C. Holmes in Sydney, Australia, the entire genetic sequence was uploaded publicly online to virological.org on 10 January, and shared with the World Health Organization (WHO), as well as with American and international infectious-disease databanks so that diagnostic tests could be invented. They gave the new virus a provisional name, 2019-nCoV.

A report of the genetic sequence was published online in the prestigious American journal, *New England Journal of Medicine* on 24 January (lead author N. Zhu). Within less than a year, that report had been cited more than five thousand times. On the same day, an article published in the leading British journal the *Lancet* (first author C. Huang) described the clinical features of the new disease caused by 2019-nCoV. It also rolled back the onset, reporting that the earliest confirmed case, to date, was a man, with no connection to the Huanan Seafood Market, whose symptoms appeared on 1 December 2019. In the space of a year, that paper was cited more than ten thousand times.

On 29 January 2020, thirty-four Chinese scientists and one Australian, published in the *Lancet* (online) the sequenced genome, together with a PCR (polymerase chain reaction) test to identify it. This rapid scientific sharing stands in sharp contrast to the ongoing political clampdown on the Chinese people and journalists who had venerated whistleblowers Li and Ai. Sharing the warnings earlier would have saved many lives. Xenophobic commentary outside China about its authoritarian response spilled over to false accusations that Zhang's lab had been closed, or that the gap between 5 and 11 January represented suppression.

This contrast between science and politics will be a recurring theme in this tale; scapegoating, another. The urge to blame someone else, often an enemy or a political rival, for every new epidemic disease has been a constant from antiquity to the present (chapters 4 and 6).

IT SPREADS

On 31 December 2019, thanks to the Chinese reports on ProMED, news of the atypical pneumonia was picked up by major newspapers. Reuters and *The Times* of London described the mysterious outbreak and reminded readers how the WHO had scolded China for its cover-up of SARS back in 2003. Toronto's *Globe and Mail* published a Reuters report of pneumonia in China on 31 December 2019, the *New York Times* on 6 January, *Al Jazeera* on 9 January, and the Russian news agency TASS on 13 January. Canadians may have been more alert and alarmed than others owing to vivid memories of their own nasty experience with SARS, which had killed forty-four citizens and caused hospitals and care homes to be quarantined for weeks on end. Word of the new outbreak seemed to be spreading faster than the highly contagious virus itself.

Since 1 January, WHO had been seeking and receiving information about the cluster of cases in Wuhan. Between 10 and 12 January, it released a set of guidelines with advice on how to manage the risk of the still unknown virus and how to conduct surveillance.

Once human-to-human spread was finally accepted by Chinese authorities (circa 20 January), the entire city of Wuhan, with eleven million people, was placed in tight lockdown beginning 23 January. No public transport, no schools, no shops, no flights. Streets were empty and silent. Travel beyond the city limits was severely restricted. The WHO labelled it an "unprecedented" move, although historians could cite examples from a more distant past when entire cities had been closed; however, the number of people involved now was indeed unprecedented. At the end of the day on 24 January, the lockdown extended to the entire Hubei province with a population of fifty million. Nevertheless, over the next week, cases appeared in every province of China, in Hong Kong, and across oceans. On dark January evenings, as numbers of cases and deaths increased, the confined citizens of Wuhan took to cheering for the health-care workers from their windows and balconies in a remarkable show of solidarity.

Uncertainty tracked the virus wherever it went. Chinese doctors soon learned the clinical dimensions of the new disease. Fever, coughing, and difficulty breathing, sometimes requiring oxygen or respirators, were joined by other symptoms, such as sore throat, fatigue, diarrhea, muscle aches, renal failure, bleeding disorders, and loss of the senses of smell and taste. The most seriously ill developed failure of many organs as their immune systems went into overdrive attacking the patient's own tissues, a situation called cytokine storm. The incubation period, from infection to symptoms, was also unclear at first, and seemed variable, from one to two weeks. The length of that period would determine the necessary duration of quarantine and isolation. The mortality rate of those who were sick appeared to be about 3 or 4 per cent, but infections could be counted only in those with symptoms. What if people could have the virus without symptoms? Unlike the influenza pandemic of a century earlier, which had been more deadly for young adults, this new disease seemed most dangerous for older people and those with other health conditions, such as diabetes, high blood pressure, obesity, or lung disease. It was culling the elderly and the chronically ill. Chinese doctors published the information quickly in open-access journals.

By 27 January 2020, the Toronto *Globe and Mail* was blaming the spread of the virus within China on "weeks of inaction." In its zeal to identify a culprit, it ignored the scientific work to identify the virus and share the information. It also ignored the mesmerizing construction activity in Wuhan as seven thousand dedicated workers, wielding multiple machines, completed the entirely new, 1,600-bed Huoshenshan shelter hospital in just ten days, ending 2 February. A second hospital, Leishenshan, based on identical plans, opened on 8 February. These shelter hospitals, called Fangcang, eventually multiplied to a total of sixteen, with twenty thousand beds. They were key to disease management: the sick did not need to isolate at home exposing others.

With its announcement of Dr Li's death on 7 February, the BBC published maps tracking the spread throughout China and Hong Kong (figure 1.1). These maps are accurate for the region, but they are deceptive. By 31 January when every part of China had seen cases, the virus had already spread to many countries around the world: Thailand, Vietnam, Japan, Korea, India, Australia, Malaysia, France, Germany, Finland, Russia, the

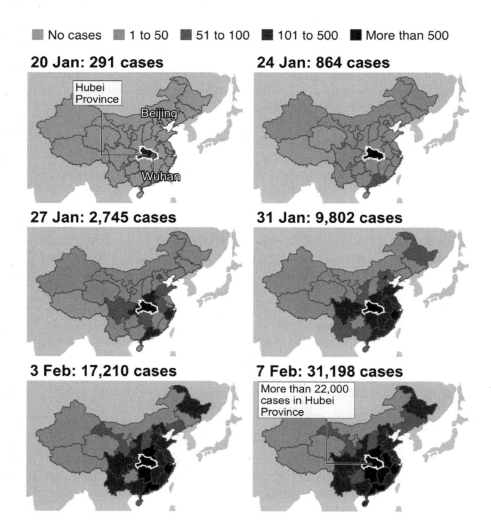

Figure 1.1 | How the virus spread in China to February 2020
(Source: China National Health Commission, from bbc.co.uk/news,
7 February 2020; accessed 2 December 2020.)

United States, and Canada, among others. The first case identified in Spain came on 31 January in a German tourist visiting the Canary Islands after he had encountered people who had visited China.

The first American diagnosed with the disease was a thirty-five-year-old man who had returned to his home in Washington state from visiting family in Wuhan on 15 January. He developed symptoms on 19 January, put on a mask, consulted a doctor, and was hospitalized. On 20 January, the Centers for Disease Control (CDC) were able to confirm with a PCR test, that he had the novel virus. On 31 January he was still in hospital and his case was published in the *New England Journal of Medicine*.

Canada's first case was identified on 25 January in a Toronto man who had also recently returned from Wuhan; the next day, his wife fell ill too. Confirmation of the new virus came two days later with the help of the National Microbiology Lab in Winnipeg. The couple was treated in isolation at their home. Soon cases were reported in British Columbia and Ontario, all in people returning from China or their close contacts. At first, these were presumptive cases because no tests were available (chapter 5). Once they could be tested, the infection was confirmed. Public-health officials watched for community spread, finding none until early March, by which time the disease had also appeared in Alberta and Quebec. My university held a public information meeting on 30 January: "The Wuhan Coronavirus: Have the Lessons Learned from SARS Helped?" The room was packed with local doctors, students, professors, and anxious citizens.

EVACUATIONS

By 1 February, Ottawa along with many other global capitals, had applied for China's permission to evacuate almost two hundred Canadians from the now locked-down Wuhan. The right to return home clashed with the Chinese right (and need) to maintain quarantine. Diplomatic negotiations were intense. Canada's situation was further complicated by the ongoing Vancouver house arrest, since December 2018, of Huawei business executive Meng Wanzhou; her confinement honoured the extradition request of the United States over the internet giant's alleged illegal dealings with Iran. In retaliation, later the same month and with no apparent evidence, China had detained two Canadian businessmen, Michael Kovrig and Michael Spavor, for alleged espionage. Tensions had already spilled over into trade relations.

Canada chartered flights to evacuate its citizens from Hubei province and authorized embassy employees to leave Beijing. The first flight landed in Trenton, Ontario, on 7 February; the second on 11 February. Other countries, including the United States, helped Canadians to leave. Evacuated citizens were to quarantine for fourteen days to avoid spreading the virus at home.

This scenario was repeated multiple times for hundreds of people from other privileged nations who lived and worked in China: France, England, Germany, Australia, New Zealand, Japan, Korea, Indonesia, Philippines, and the United States. Families were separated as the limited seats on flights forced difficult decisions about sending children home and leaving behind a working parent. Those arriving home were ordered to quarantine, either in their own dwellings or in hotels. Some evacuees were housed in gilded-cage resorts in southeast France or Phuket, Thailand. Repatriated Canadians from Wuhan were accommodated more modestly at the Trenton military base.

Notwithstanding the heavy precautions taken for evacuees in early February, the infection had already spread widely and rapidly through travellers leaving China before the lockdown and before the evacuations.

CRUISE SHIPS

On 4 February, the media was alerted by passengers to an outbreak of ten positive cases on the luxury cruise ship, *Diamond Princess*, sailing in Japanese waters. The ship was quarantined in the port of Yokohama and passengers were forced to remain on board. The first case in this cluster might have been an eighty-year-old man who had boarded in Yokohama on 20 January and disembarked five days later in Hong Kong where he fell ill and was hospitalized on 31 January. Passengers were confined to their cabins and meals brought to them. Many complained about the ship's poor ventilation and sloppy sanitary measures, fearing that their isolation meant greater exposure. Japan had initially been reluctant to let anyone off the ship. Once again, countries stepped forward to arrange evacuation to shore for their stranded citizens. Eventually the sick were treated in Yokohama hospitals. Canada chartered another jet to bring 129 passengers into quarantine at a hotel in Cornwall, Ontario. By 1 March when all had left the ship, more than seven hundred of the 3,700 people on board had tested positive for the virus, and fourteen had died.

The *Diamond Princess* became an epidemiological experiment in a closed population that shed light on areas of clinical uncertainty. Almost everyone on board had been tested, some more than once. The results allowed researchers to understand that 18 per cent of people infected with coronavirus had experienced no symptoms at all; yet they could infect others. This insight meant that many early cases in China had likely gone undetected and the mortality rate there might be lower than originally measured. The results also confirmed the greater risk to the elderly and the easy transmission of the very contagious virus within the close quarters.

Nevertheless, the cruise ships kept sailing. By April, outbreaks had happened on fifty-five different vessels and killed at least sixty-five people, though precise numbers were (and still are) difficult to obtain. One such vessel was the *Grand Princess* ordered into quarantine on 4 March off the coast of California with more than 3,500 passengers from fifty-four countries. That same day saw the first California death: a man who had been a *Grand Princess* passenger on its earlier voyage to Mexico between 11 and 21 February. *Grand Princess* was said to have been a breeding ground for the disease. President Trump was reluctant to allow the ship to dock. Beginning 9 March with at least twenty-one people already infected and fear of greater spread on board, the passengers were slowly taken off the ship over the course of three days and moved into the port of Oakland, where they were quarantined on nearby military bases or sent home. Canada chartered another jet to bring home 228 passengers for a fourteen-day quarantine at the Trenton military base that had only recently been vacated by the Wuhan evacuees; eventually at least thirteen tested positive.

People of modest means had invested their savings in those long-awaited dream vacations, only to have the palatial ship become a prison, and their voyage, a nightmare. Final numbers of infections and deaths from the *Grand Princess* and other cruise-ship outbreaks are unknown, and dozens of lawsuits for negligence were filed against the owners.

THE WORLD HEALTH ORGANIZATION (WHO)

Created in 1948 following World War II, the WHO is the United Nations agency responsible for monitoring and improving global health. It has been a steady source of information, vigilance, and policy guidance for more than seventy years. Its original goals included improving quality and length of life and eradication of malaria and smallpox. Its first director general was Canadian military psychiatrist, Brock Chisholm.

In 2020, the WHO had 194 member states and its director general was Dr Tedros Adhanom Ghebreyesus who had been serving since 2017. An Ethiopian-born expert in community health and infectious disease,

Tedros had studied in the United Kingdom, defending his 2000 PhD thesis on the effect of dams on malaria transmission in his homeland. He had participated in numerous international programs against malaria, tuberculosis, AIDS, and childhood diseases. His WHO agenda included advocating for universal health coverage and attention to the environment and climate change.

The WHO had been watching the Wuhan situation since the first report. By mid-January, it had tracked the spread, convened meetings, and issued guidance documents for monitoring and isolating suspected cases of the new disease. It urged caution, disseminated scientific information, and warned countries to prepare for the advent of the contagion. In particular, the WHO pointed to the importance of keeping the rate of infections low to avoid overwhelming health-care resources (table 1.1).

On 11 February, the WHO announced a name change for the new virus, from "2019-nCoV" to "SARS-COV-2" – a decision taken to correspond to standards of nomenclature. That same day it chose the name "COVID-19" for the disease (COronaVIrus Disease of 2019). The name was created deliberately to avoid the typical blaming that is heaped on places or creatures when outbreaks were known by their geographic origin or by animals that shared the pathogen. On 11 March with 118,000 cases in 114 countries and four thousand dead, Dr Tedros declared that COVID-19 had become a pandemic.

The WHO referred to the importance of *"flattening the curve,"* a vivid expression that its experts had used since at least 2007 in confronting influenza. Maybe the same total number of people would eventually be infected, but survival would be much greater if people did not all fall ill at the same time overwhelming medical capacity to deal with them.

One of the obstacles to the work of the WHO has always been pushback against what is perceived to be unwelcome, foreign interference. In fact, suspicion between nations had effectively blocked the nineteenth-century Sanitary Conferences that had attempted, on multiple occasions since 1851, to create international health cooperation without success. Not only did countries resent outsider intrusion into their policies, they rejected the idea of notifying each other about public-health problems for fear of making themselves vulnerable militarily, politically, and economically.

Criticisms of WHO recommendations are not new; they go with the territory of such work. In 2003, Canada watched in embarrassed horror as Toronto's mayor, Mel Lastman, famously railed against the WHO for having issued a travel advisory during the SARS outbreak. Furthermore,

Table 1.1 | Timeline of some early WHO actions, 2020

1–5 January	Gathers information about the Wuhan situation.
10–12 January	Issues comprehensive guidance for managing an outbreak.
19–21 January	Issues statements about the human-to-human transmission.
26 January	Issues a free online course about the new virus.
30 January	Declares a Public Health Emergency of International Concern (PHEIC) in nineteen countries.
11 Februry	Announces disease name will be COVID-19; its viral cause is renamed SARS-CoV-2.
11–12 February	Convenes a global forum with three hundred experts from forty-eight countries.
11 March	Declares COVID-19 to be a pandemic (in 114 countries; four thousand deaths).

its well-intentioned policies can backfire. In the H1N1 influenza epidemic of 2009, which began in Mexico, the WHO assiduously avoided the previous practice of naming diseases after the location of the first cases, because the stigma could result in economic harm. Instead, it initially chose to name that H1N1 outbreak after the animal that most commonly shares our influenza viruses: the pig, even though no animal-to-human spread had been documented. The swine flu label inadvertently led to trade embargos on pork and the futile slaughter of thousands of innocent pigs turned into hapless scapegoats. Neither scientists nor the WHO had recommended such actions, and they were powerless to prevent them. Countries acted alone simply because of the chosen name and the fear it inspired. Also, some contended that the WHO's declaration of the H1N1 pandemic had been premature and prompted by pharmaceutical companies. Possibly in reaction to these charges, the WHO response to the 2014 Ebola outbreak was criticized for its slowness, complacency over tepid local reports, and lack of coordination.

Likewise, in 2020, the WHO faced another barrage of disapproval in its response to the new coronavirus. As early as 23 January, it was criticized for not declaring a pandemic. After Dr Tedros and a WHO delegation visited Beijing in late January, it was criticized again, especially by then President Trump, for being manipulated by China. It was also criticized by the scientists who had chosen the original viral name for supporting the change to SARS-COV-2. WHO caution over declaring the pandemic and its choice of the anodyne disease name – devoid of geography and animals – were surely informed by the criticisms and lessons of 2009. Nevertheless, the human proclivity to blame the other and label suspicious sources was relentless: as variant strains emerged later in 2020 and 2021, location names stuck to them like glue: UK, South Africa, Nigeria, Brazil, India.

The WHO also had to combat a campaign of misinformation, disinformation, and conspiracy theories. The harried Dr Tedros labelled it an "infodemic," in a Munich speech on 15 February 2020. The extent of this parallel, multination, digital scourge only grew with the passage of time (chapter 10). It originated with suspicious individuals only to be ramped up by the geopolitical aspirations of state leaders bellowing lies into resounding echo chambers. The parallel infodemic made the pandemic even more difficult to control.

BACK TO CHINA, ON TO EUROPE, AND THE GLOBE

As COVID-19 spread around the world, China's numbers grew, peaked in mid-February 2020, and declined. The total case count, at just under ninety thousand and deaths at 4,500, seemed staggering at the time; but it was a pale reflection of what was to come elsewhere. Despite the falling case numbers, Chinese officials maintained the tight lockdown for several more weeks. By mid-March, they cautiously began lifting restrictions on Hubei province. Complete reopening was declared on 8 April to the boundless joy of its population. On that day, the seven-day average of new cases had been fewer than one hundred for an entire week.

From the beginning, experts were warning of a second wave. Indeed a few hundred cases appeared in China in July, but they occasioned only a handful of deaths. China did not have a second wave until early 2021. As the case rate exploded in other parts of the world, its relatively good outcome occasioned much analytic chatter. It was said that these chronically oppressed people were docile, accustomed to obeying orders; they were used to wearing masks; they were subservient to their economy.

Resistant to crediting the remarkable result achieved when little about the pathogen was known, western observers, smug in their ideals of liberty, dismissed China's success: it was not something to emulate; it was a manifestation of totalitarian rigidity; it might not even be true.

Yet there was no avoiding the fact that the people of Wuhan had stopped the spread of the virus, using the ancient methods of quarantine, isolation, and hygiene – without specific drugs and without vaccines. Later analysis suggested that had the interventions been launched one week earlier, more than half the cases could have been prevented, but it also demonstrated had they been launched a week later, the caseload would have tripled. Horrifying as the Chinese numbers had been in early 2020, that nation of 1.4 billion people has one of the lowest case numbers per capita in the world.

Meanwhile, after belittling the seriousness of the disease and failing to warn his citizens abroad, the president of the United States blamed China for the virus and its spread. He suspended all flights from that country on 11 March, well after Hubei's peak in cases and the same day that WHO declared a pandemic. He tilted against the scientific name and persisted in referring to it as the "Chinese virus," doing so publicly more than twenty times in the latter half of March alone. Trump's rhetoric and actions left many Americans surprised to be stranded in Asia and sparked remarkably resilient, conspiracy theories that it was a weaponized virus escaped from a Chinese laboratory or that effective treatments were being suppressed.

Europe became the new locus of COVID-19 infection. Following sporadic cases in citizens returning from China during January and February, community spread by March and into April, led to hundreds, and then thousands, of new cases daily in Italy, France, Germany, Spain, Sweden, Poland, the United Kingdom, and other countries. They were unevenly distributed across various regions. Italy was hit hard in the northern regions of Lombardy and Piedmont surrounding the great commercial cities of Milan and Turin. France saw disastrous caseloads in the northeast. Spain was most affected in its centre, around Madrid.

An evangelical gathering, between 18 and 21 February, in Mulhouse, a small city in eastern France, had assembled more than a thousand people from around the country and beyond. It was eventually recognized as a superspreader event that provoked at least 2,500 infections on three continents. A similar gathering in South Korea was implicated in five thousand cases. These were just two of many religious assemblies that resulted in leaders of every faith ordering in-person ceremonies to stop.

These public-health decisions sparked immediate backlash from the devout who claimed that the state had no business messing with their unalienable right to practise their religion.

Hospitals and ICUs were overwhelmed. All but the most urgent surgeries were cancelled. Health-care workers sickened or collapsed with infection or burnout from the burden of their work. Seriously ill patients were transferred to different regions and even different countries. France transformed a TGV railway car into a moving ICU to convey patients from crowded hospitals to others with available beds. The military was invoked to set up field hospitals in parking lots. Corpses accumulated in morgues and hallways. And the media began to take stock of the shocking mortality in care homes for the aged and disabled.

Also in February 2020, the disease was reported in Iran. Following its first case on 19 February, the state quickly intervened to close schools, shrines, shops, secular gatherings, and Friday prayers. Iran hesitated to shut down entire cities, although individuals were ordered to isolate and quarantine, and travel was restricted. Already suffering from trade sanctions imposed by the United States, its leaders praised citizens for cooperating and ordered the military to help clean streets and work in clinics. During March, more than two hundred thousand prisoners were slowly released to avoid the spread in prisons. Riots for more protection ensued among the retained prisoners of conscience and so-called security cases, some of whom were killed in the violence.

For those of us watching the spread, one country seemed to remain suspiciously untouched, although it shared a long border and close ties with China. Russia did not begin reporting significant cases until well into March when Europe was deeply affected, and China's numbers had declined. Nevertheless, on 20 January, the Kremlin-based media outlet, *Zveda*, quoted one Igor Nikulin, a man of obscure credentials, who suggested that the virus had been engineered by the United States – an idea directly opposing claims of the American president. Both theories entered the infodemic.

By mid-March community spread was evident in both North and South America too. At the end of the month, Iran continued to simmer as the latest hot spot, and the virus flowed between the Middle East, Europe, and Russia with travellers who had never dreamt of going to China.

2

COVID-19 Comes to North America

We knew it was coming. Uncomfortable as it may have been, the evidence gave no reason to believe that any country would be spared. So much of what happened next was a performance of reaction, instead of a strategy of prevention.

The first confirmed cases in the United States and Canada were identified on 21 and 25 January, respectively. Seattle reported the first American death on 29 February and two days later came a death in New York state. Soon after those events, New York City became the centre of a massive outbreak of COVID-19, which had yet to be declared a pandemic.

On 7 March, New York Governor Andrew Cuomo declared a state of emergency and began issuing guidelines across the following weeks – guidelines that tilted toward an increasingly tight lockdown. One-by-one, he closed schools, bars, restaurants, nonessential businesses, and construction sites. He banned large gatherings, issued work-from-home orders, and placed a nighttime curfew on the subway, which deepened hardship for workers in essential services. By 27 March, the United States had seen more COVID-19 than any country in the world. Two weeks later, New York state had more cases than all the other American states combined.

New York City marked one thousand total deaths in late March, and by early April, more than eight hundred people were dying each day – four times the city's usual death rate. The burgeoning numbers of corpses soon exhausted the capacity to deal with them. The city brought in refrigerated transport trucks to store the bodies. The unclaimed dead were laid in plain, pine boxes and buried in mass graves at the old potter's field on Hart Island in Long Island Sound. Over the next four months, almost 23,000 people died of COVID-19 in that city alone.

Not until 15 April were people in New York state ordered to wear face masks in crowded, public spaces. The delay of that recommendation is emblematic of cultural aversion to masks and confusion over their utility (chapter 6). During this crisis, squabbling arose between the city's public-health officials and the mayor, adding to the chronic strain between the mayor and the governor. The tensions were about preserving the economy versus preserving safety and survival. They were deepened by attitudes to authority, ranging from full respect, through dubious suspicion, to outright opposition.

New York state's total number of active cases soon ceded its top spot to others, including California, Texas, Florida, and Illinois. But its case fatality rate, with those of New Jersey and Rhode Island, remained high at about 5 per cent. Two reasons in particular stand out to explain why. In the first wave, testing was sporadic and incomplete: the confirmed case count did not include all those infected, nor did it include people without symptoms. The true case numbers may have been much higher and proportion of deaths, therefore, lower than reported. Also, the number of sick people rose so quickly that health services were completely overwhelmed; not everyone who needed life-saving support could receive it. All by itself, New York state's death rate proclaimed the importance of flattening the curve.

Philadelphia also experienced many cases of COVID-19. To learn more about the course of the disease, doctors at Temple University began collecting data from 513 patients admitted over five weeks from 10 March. They were interested in a better description of the disease and especially in identifying factors that could predict which patients might develop the dreaded outcome of cytokine storm. That immunological meltdown had first been recognized and named by transplantation scientists back in the early 1990s, especially when bone-marrow grafts rejected the host patient. It had also been seen in SARS and in MERS. The doctors eventually introduced simple laboratory criteria that could herald more severe illness: changes in white cell counts, enzyme levels, and electrolytes. The analysis was finally published in November 2020.

On a national level, the looming health crisis provoked conflicting messages. On one hand, President Donald Trump, campaigning for re-election, tried to ignore it, resorting to anodyne quips, lies, and facile posturing. Canadian journalist Daniel Dale, working at CNN, and large media entities, including CBS News, the *New York Times*, the *Washington Post*, and *The Atlantic*, constructed factchecks and timelines of his numerous, inaccurate statements about COVID-19. Here are just a few examples garnered from the president's tweets and interviews:

22 January	"We have it totally under control." (Repeated often until 29 February.)
26 February	"This is like a flu."
27 February	"One day, it's like a miracle, it's going to disappear." (Multiple times between February and October.)
6 March	"Anybody that wants a test can get a test."
10 March	"We're prepared and we're doing a great job."
23 March	"Suicides [provoked by economic shutdown] ... definitely would be in far greater numbers than the numbers that we're talking about with regard to the virus."
4 July	"99 per cent [of COVID-19 cases] are totally harmless."

At the outset, the president complimented China's efforts to quickly identify and control the pathogen. But by late March, he had switched to military rhetoric of a battle, repeatedly referring to the "Chinese virus," over the outraged objections of Asian Americans who were being targeted by xenophobic attacks. Later he blamed Mexicans.

On the other hand, public-health officials expressed concern and urged citizens to follow rules of hygiene. In particular, the highly respected, infectious disease expert Dr Anthony Fauci, of the National Institutes of Health, advocated social distancing and warned that the virus was ten times more deadly than influenza and twice as contagious. He admitted that implementing mitigating precautions earlier would have saved lives. For his honesty, certain Republicans called for his ouster. Eventually he and his family needed a security detail because of death threats. Fauci and the Food and Drug Administration director, oncologist Stephen Hahn (appointed in 2019), were often called upon to corroborate the president's statements. Their resistance was obvious, their discomfiture, palpable.

By March, many Americans were out of work because of the pandemic and measures taken to control it. Stock markets plummeted – temporarily, as it turned out, but that could not be known at the time. With restrictions on travel, observers predicted that airlines would go bankrupt. More than three million Americans filed for unemployment relief in the week of 21 March. Following a few, feeble attempts at offering financial support earlier in March, the massive *Coronavirus Aid, Relief, and Economic Security Act* (CARES) and its relief package of over US$2 trillion, was signed into law on 27 March, with remarkable bipartisan support in its crafting and having passed the Senate unanimously. Pundits contrasted Republicans' support for this measure with their resistance to relief packages during the recession of 2008–09, speculating on whether (or not) their participation would have been forthcoming had the POTUS been a Democrat. By May, US$10 billion of the CARES package was directed to Operation Warp Speed, a public-private partnership to develop vaccines and treatments to fight the pandemic.

By 14 April, having finally accepted that COVID-19 was a real danger, Trump was blaming the WHO. In the midst of this global catastrophe, he announced that he would cut American funding to the WHO, an act that *Lancet* editor, Richard Horton labeled "a crime against humanity."

The president basked in his announcements over CARES and the WHO, but he famously flouted all hygiene recommendations, refusing to wear a mask and insisting on hugging or shaking hands with prominent leaders. By example, he compelled his entourage and acolytes to do the same, and he held massive rallies with unmasked citizens across the country – rallies that became superspreader events (figure 2.1). In early June, he used tear gas and inordinate force to dispel a crowd peacefully protesting police brutality, so that he – an unrepentant liar, an adulterer, a maker of thieving deals – could stand in front of a church for a photo op holding a Bible that was not his own.

Smoldering for several years before the pandemic was the rising anger and protests of the Black Lives Matter (BLM) movement with its rage against the racist cruelty of police. The movement exploded in 2020 with the police-shooting deaths of Breonna Taylor on 13 March and George Floyd on 25 May, and the maiming of Jacob Blake on 23 August. More than ten thousand protest rallies, marches, and strikes were held across the country during the summer of 2020. Some were accompanied by violence.

The mayor of Washington renamed two blocks of a street near the White House, Black Lives Matter Plaza; it had been the site of Trump's biblical photo op. Some used the pandemic as a reason to oppose the actions of BLM and cried hypocrisy when Trump rallies were said to be dangerous and those of BLM were not. Experts eventually concluded that BLM contributed little to the spread of infection because protesters understood the danger: they wore masks, took protective measures, and were usually moving outside in the sun and wind, rather than standing shoulder-to-shoulder, bellowing, maskless, in packed stadiums. When BLM protestors went home, some quarantined. The big difference between the groups was belief in the risk, but this was a public-health message that went underappreciated.

Figure 2.1 | Editorial cartoon by Bruce MacKinnon, *Chronicle Herald*, Halifax, 12 June 2020

When Bob Woodward's book *Rage* appeared on 11 September 2020, reviewers focused on the author's description of a 7 February interview with Trump, during which the president admitted that he had been fully aware of the coming danger. He justified his early pandemic lies by insisting that he had wanted to avoid creating panic. Two weeks later, almost in defiance, Trump presided over a closely seated gathering of two hundred, largely unmasked people in the White House Rose Garden to nominate Amy Coney Barrett to the Supreme Court. Her rushed nomination was a flagrant political ploy to stack the court with conservative judges before the election. For public-health officials, it too became a superspreader event, a never event, a disaster that did not need to happen. By early October, more than a dozen guests had been infected, including the president himself, his wife, and son.

As much as Trump played a prominent role in the chaotic American response to COVID-19, the nation's public-health system had gradually been starved for funds dropping from 45 per cent of the federal health-care budget in 1960 to only 15 per cent. The states were to pick up the slack, but only some did. Consequently, national rules and policies were feeble. Furthermore, a federal plan for health emergency preparedness had been cut by more than three hundred million dollars between 2002 and 2017. Over the same period, the United States spent proportionately less on prevention than either Canada or the United Kingdom, while according to the Organisation for Economic Co-operation and Development (OECD), its rate of preventable deaths was higher than all comparable counties.

CANADA

As the sordid drama described above was unfolding in the United States, Canada was experiencing its own first contacts with the new virus. After a Toronto couple became the nation's first confirmed COVID-19 cases on 25 January, sporadic infections were identified in people who had recently travelled to Asia. On 12 February, Ontario reported that a third case had resolved.

British Columbia became the initial focus of the first wave. With its sixth case, on 20 February, came the announcement that the traveller had been to Iran, not China. Iran was also implicated as the origin of Quebec's first case, while Toronto reported the disease in a traveller from Egypt. British Columba also announced the country's first case of community spread (5 March) and its first COVID-19 death (9 March). Three days later, Manitoba, Saskatchewan, and New Brunswick had reported cases.

On 12 March, the country learned that Sophie Grégoire Trudeau, wife of the prime minister, had tested positive for COVID-19, having flown home from a speaking engagement in the United Kingdom. The prime minister announced that he too would be in isolation for the fourteen-day quarantine period. Nevertheless, he felt well, continued working, and gave daily press conferences from the steps of his home to journalists assembled under a cloth gazebo. Both Trudeaus urged Canadians to follow public-health advice.

By mutual agreement, the border between Canada and the United States closed for nonessential travel at midnight on Friday 20 March. The closure was to last thirty days. Little did the two countries realize that it would be extended multiple times and well into the following year. Trucks with provisions, such as food and medicine, would be allowed through both ways.

Unlike the constant display of friction between American politicians and public-health officials, Canadians grew accustomed to the press conferences of premiers, standing with their ministers of health and often accompanied by medical officers of health and others who presided over long-term care. The federal government did the same. Out of isolation and free of infection, the prime minister rejoined the federal team at press conferences. They stood (or sat), solemn and socially distanced, with flags prominently displayed, as team members delivered their news one by one. The subliminal message was solidarity in a united front.

For me, an old doctor who had graduated in the very different world of the mid-1970s, watching the array of public-health officers leading the pandemic response was exhilarating. So many of them were female! And, notwithstanding their gender, they quickly became heroes. When I entered the profession, many women already worked in medicine, but few were chosen to lead. For example, no woman had ever served as a dean of academic medicine in Canada until 1999. I had not noticed when the leadership in public health had grown gender blind.

These physicians were led federally by the diminutive, plain-spoken Dr Theresa Tam. Born in Hong Kong, raised and educated in the United Kingdom, Tam had specialized in pediatrics at the University of Alberta and in infectious diseases at the University of British Columbia. A public servant for many years, she had been immersed in Canada's response to SARS, following which she co-authored a 2006 report that laid out plans for managing an anticipated respiratory pandemic. The position that she occupies was created as part of pandemic planning in the wake of SARS. Her qualifications were impeccable.

By April, Dr Tam was the brunt of racist and misogynist slurs from citizens across the county, including Conservative Member of Parliament Derek Sloan, campaigning to lead his party. He alleged that she was blindly following the WHO and the Chinese government. He refused to apologize and lost his leadership bid on the first ballot. Protests over his behaviour from Chinese Canadians, the LGBTQ community, women's groups, and pro-choice activists had little effect; only in early 2021, after it was revealed that he had enjoyed campaign donations from a white supremacist, was he expelled from the party. Sloan sat in Parliament as an independent for a rural Ontario riding, until the 2021 election when he ran in Alberta and was defeated.

While Canadian public health has federal standards, health care is administered provincially – and the country's pandemic management displayed marked differences in its ten provinces and three territories, a situation that epidemiologists, such as Kumanan Wilson, have long recognized as an unfortunate by-product of federalism, the same disparate situation that hampers American public health.

Leading British Columbia's response was Dr Bonnie Henry, a specialist in public health, whose career had also been geographically extensive. Born in Prince Edward Island, educated at Dalhousie University, with additional training in San Diego, she had served in the Canadian navy, in Toronto during SARS, in Pakistan fighting polio, and in Africa with the WHO during an Ebola fever outbreak. Canadians were riveted by her soft voice, calm demeanour, and obvious emotional connection to the grim reports. She reminded everyone to be kind, to take care of each other, to stay the course. Her quirky, designer shoes and artistic necklaces grabbed social media attention, and admiring citizens sent her homemade gifts and created a fan club on Facebook. With her sister, she published a book on the first four months of the pandemic and her earlier public-health guide was reissued. In a 5 June 2020 profile, the *New York Times* lauded the self-described introvert for her powers of communication, calling her "one of the most effective public-health officials in the world."

The focus would soon shift elsewhere, as first-wave lockdowns occurred in different provinces of Canada. One problem was the vastness of the land, which meant that a raging outbreak several thousand kilometres away did little to motivate precautions that would severely limit income and social life. As testing became more widely available, Ontario and Quebec soon took over from British Columbia in having the most cases.

On 12 March, a day after Ontario's first death, Premiers Doug Ford of Ontario and François Legault of Quebec announced the closure of all public schools for a two-week period, which ended up extending by increments to the following autumn. Teachers scrambled to develop online learning. Four days later, Ford recommended closing all daycare centres, sporting events, faith gatherings, bars, and restaurants (except for takeout). By 23 March, Ontario ordered closure of nonessential businesses, exempting realty, construction sites, and liquor stores. Remembered as the "buck-a-beer" premier, Ford and his associates argued that the last thing the overstretched hospitals needed was an outbreak of alcohol withdrawal on top of everything else.

Quebec also piled on similar lockdowns one after another. Quebec City implemented drive-through testing on 18 March, followed by facilities elsewhere. To keep hospital beds available, elective procedures for orthopedics and cancer surgeries were postponed or simply cancelled, compounding community distress. By May, case counts slowly began to fall in some areas of both provinces, and cautious reopening was permitted in graded phases, region by region.

The arrival of the pandemic reignited many of the old blaming and scapegoating behaviours against Asians, much to the dismay of public-health workers who deplored the intolerance and racism. Bonnie Henry kept calling for compassion: "Be kind!" As the weeks and months rolled by, the initial affection for her and her colleagues wore thin, and they too endured insults and even death threats.

The federal government was on the receiving end of some of these attacks, especially when the country learned that its once highly respected alert system had been dismantled just a year before the pandemic. Canada's Global Public Health Intelligence Network (GPHIN, pronounced G-fin) began in the 1990s when an outbreak of pneumonic plague in India sparked concern globally and at Toronto's Pearson airport. American-born-and-trained Ronald St John set up the internet scanning system to alert the world to potential infectious dangers. Using a sophisticated approach to news reporting, it was improved following SARS in 2003 to become a reliable tool widely adopted around the world. But when no new pandemic emerged, it was slowly starved of support and finally dismantled in 2018. Critics claimed that, without it, Canada's response to the danger of the coronavirus was inordinately slow.

LONG-TERM CARE DISASTER

It was Dr Bonnie Henry's empathic and occasionally tearful briefings that forced Canada to take stock of the special vulnerability of elders and the lax standards in care homes. The country's first outbreak in a long-term care (LTC) facility had occurred at the Lynn Valley Care Centre in North Vancouver on 5 March. By mid-March visits were reduced in care homes, and staff members were limited to working in a single site. Because of low wages, these workers often held down several jobs to make ends meet. Observations from France and nearby Washington state had warned that LTC outbreaks would grow in number and extent. And grow they did.

With the comings and goings of so many staff and visitors, the virus easily invaded LTC homes, conveyed especially by those without symptoms. Once inside, the virus spread widely though caregiver contact with residents and crowded groupings for meals, activities, and even sleeping. The special vulnerability of the elderly meant that the suffering and carnage was devastating. Bonnie Henry called it a "tragedy."

By the end of March, Dr Henry had banned all visits to British Columbia's care homes and ordered staff to wear masks and gloves and follow sanitary protocols. She also ensured that the displaced workers would receive government support for lost wages. Nevertheless, the outbreaks multiplied, and deaths rose.

Other provinces knew the risks but hesitated – like deer in headlights – to implement these basic precautions. When on 14 April Ontario limited LTC workers to a single site, the disease was already in 114 of its care homes. Quebec had issued the same order on 2 April; Nova Scotia on 17 April. Methods of control were severely hampered by the initial shortages of personal protective equipment (PPE). But the pandemic held up a mirror to all Canadians, showing them that conditions in LTC homes were deplorable. As the first-wave pandemic waned, the death toll in Europe also proclaimed greater danger for the elderly, especially if they lived in care facilities. But a larger proportion of European seniors remained in their own homes or with families. Though grievously harmed, they fared better than Canadians.

Canada's elders and disabled had been abandoned long before by cultural attitudes that fostered penny-pinching public policies. Care facilities had never really been integrated into health-care funding. Standards varied across provinces – another federalism problem. Many homes were run by for-profit, private entities; government inspections had been reduced or abandoned. The advent of COVID-19 restrictions meant that family members could not visit their loved ones or offer comfort to the sick and dying. Anxious and bereaved, they demanded answers. Together with professional associations for nurses and support workers, they launched lawsuits against the owners.

So dire was the situation that on 22 April the premiers of both Ontario and Quebec invoked the military to feed, wash, and tend to patients in an eventual total of fifty-four LTC homes. What the soldiers found was a national disgrace: residents were ignored, filthy, dehydrated, and crammed into stuffy, small spaces, impossible for isolation. This aspect of the Canadian Armed Forces' extensive pandemic response, Operation Laser, eventually cost taxpayers C$53 million.

In late August, a team from McMaster University and the University of Toronto published an analysis of Ontario's LTC situation in the *Canadian Medical Association Journal*. It confirmed what many had suspected: larger, older homes, built in the 1970s or earlier, could not adapt to the demands of sanitary measures. Those run for profit had the same risk of an outbreak, but once established, outbreaks in for-profit homes were larger and more deadly, especially for entities owning chains of homes. Calls for action, revised standards, and revived inspections had already been implemented. Many finally began to wonder why Canada's supposedly universal health-care system has neglected home care and chronic care for the elderly and disabled.

At the end of the first wave, most of the seven thousand Canadian dead had been seniors in LTC. The carnage from LTC homes put that country's mortality rate from the first wave pandemic at 5 per cent of cases – one of the worst in the world. A June report released by the Canadian Institute for Health Information (CIHI) showed that while total COVID-19 deaths per capita in Canada were just below the average for OECD countries, those of LTC residents were well above average. In fact, the 81 per cent of all Canadian deaths in LTC was by far the highest of seventeen wealthy nations, the average of which was 38 per cent. The second highest was Spain with 66 per cent. Furthermore, nations that had implemented early and consistent LTC controls had better outcomes. Canada's treatment of people in care was exposed on the international stage as flagrantly uncaring.

IN THE EAST

In general, the Atlantic and prairie provinces were mostly spared large outbreaks during the first wave. However, the sense of gloom was deepened by a COVID-19 outbreak centered on Campbellton, New Brunswick, and a horrifying mass shooting in Nova Scotia on the night of 18–19 April 2020 by a gunman disguised as a police officer. The Campbellton outbreak was small by standards elsewhere, but it exposed a nasty desire to blame, combined with ugly, racist attitudes.

Dr Jean-Robert Ngola, born in Congo, educated in Belgium, had been working in Canada since 2005. Provincial guidelines would normally require a fourteen-day quarantine for those coming from out of province. But Campbellton lies on the border with Quebec, and many were permitted to cross both ways daily for work or supplies. On 12 May, Dr Ngola made the nine-hour drive to Quebec City to pick up his four-year-old daughter so that her mother could attend a funeral overseas. Before leaving, he contacted police and was assured that, as a health-care worker, he would be exempt from quarantine and could return to work using precautions. On 25 May 2020, the Campbellton outbreak was declared in a hospital patient. The doctor and his daughter were among forty people who tested positive, two of whom eventually died. Before contact tracing was complete, Quebec was deemed the source, and Dr Ngola was publicly identified, by Premier Blaine Higgs and others, as patient zero. He was fired without pay and his licence to practise medicine was suspended. There followed racist taunts and physical threats, and he had to relocate for his safety. Ngola had probably contracted the virus in Campbellton after his journey; yet, he was charged in August 2020 with violating the *Emergency Measures Act*, despite support from as many as 1,500 doctors nation-wide. The case was not slated for trial until June 2021. The unfair charges were finally withdrawn only days before the court date. Despite repeated urging from Ngola's lawyers and supporters, Premier Higgs rejected calls for an apology.

The ugly scapegoating and blaming continued against racialized peoples and against those who, for whatever reason, tested positive. It was a sad human tradition: Jews had been blamed during bubonic plague; immigrants and the poor during cholera; Spain during 1918 influenza; Haitians and homosexuals during AIDS; Asians during SARS. In COVID-19, the social stigma affixed to performing risky work or testing positive was sometimes sufficient to lead people to avoid testing or hide its results, provoking more danger. Some of those injured by intolerance, like Dr Ngola, were forced to leave their homes. Surprisingly, Manitoba ethicist Arthur Schafer contended that public shaming – naming and blaming – should become a tool in the public-health arsenal, opining that it would be more effective at altering behaviours than stiff fines.

Notwithstanding the widely publicized Campbellton outbreak of late May, the generally low-case rate in the East, prompted the four eastern provinces to band together, creating on 24 June 2020 what became known as the Atlantic Bubble. In many ways, their region was like an island – Prince Edward Island and Newfoundland were indeed islands – while the number of overland points of access were limited, making it relatively easy to control the land borders with the rest of Canada and the United States. From 3 July, the Bubble allowed unrestricted travel between the four provinces. Anyone entering from outside would be screened for symptoms and required to quarantine for fourteen days. For a time, it attracted praise and envy from other lands.

THE (LOST) WAGES OF THE PANDEMIC

In all nations, considerable hardship stemmed from the need to close businesses and schools. Remote work – *télétravail* – and online schooling attempted to allow economies and education to proceed apace. But the transition was not easy. Workers yearned for the stimulation and society of their colleagues, and they resented attempts by employers to spy on their computers, ensuring time commitment to their jobs. Teachers and children, both, struggled with the online classes. Parents – mothers especially – had to take on additional responsibilities to help their offspring with the adjustment, while juggling virtual work themselves in addition to caring for even younger children whose daycare facilities had been closed. These problems were aggravated when income collapsed, jobs disappeared, and landlords demanded rent. Depression and anxiety rose.

To allay financial distress, on 18 March 2020, the Canadian government announced an C$82 billion aid package that contained several envelopes: increased child benefits; flexibility on repayment of income taxes and loans; doubled rebates on the Goods-and-Services Tax (GST); and the Canadian Emergency Response Benefit (CERB) that would pay $2,000 monthly to those who had lost jobs through pandemic-related closures, quarantines, and illness. The latter came into effect on 6 April; by the end of that week, more than 3.5 million people had applied. More help came on 11 April with a wage subsidy for eligible companies affected by lost business to help maintain employment and income.

A planned $900 million emergency package for postsecondary students, announced on 22 April, provoked a political crisis. Federal employees were working under difficult circumstances online and from home, as they also coped without school or daycare for children. The vast number of new programs overwhelmed their ability to administer applications and payments. Consequently, the government appealed to a charity to provide oversight for administering the student package: the WE Charity, founded in 1995 as Free the Children and run by human-rights activists Craig and Marc Kielburger. Because Trudeau family members and the finance minister had helped this charity in the past (and received honoraria), the opposition parties accused the government of favouritism and demanded an investigation by the Office of the Auditor General of Canada (OAG) and by the office of the Conflict of Interest and Ethics Commissioner of Canada. The choice had been made because of the charity's extensive administrative facilities and, crucially, without input from the prime minister's office; however, the fallout resulted in cancelling the contract (July) and the eventual resignation of the finance minister (August). It also led to wider revelation of the charity's unsavory, cult-like activities, probable fraud, and disgrace of the Kielburgers who complained that they had been scapegoated by politicians, while their families were subjected to death threats. They closed operations in Canada. For some Canadians, the opposition's milking of the prime minister's scandalous involvement with a charity (however suspect), from which he received no benefit, made a ludicrous contrast with the shocking scandals and reckless behaviour of the president south of the border. It became a tedious, noisy distraction in the middle of a mounting crisis. In May 2021, the office of the Ethics Commissioner eventually cleared Trudeau of any involvement in the choice of the charity, but the promised financial relief for students never appeared.

COVID-19 revealed longstanding inequities. It differentially affected those who could not conduct their work from home. Low-wage essential workers – in warehouses, packing plants, abattoirs, and garbage collection – lost wages when they fell ill or were obliged to quarantine. Many service workers, often women in retail and restaurants, lost their jobs when businesses were forced to close.

Moreover, at least 58 per cent of Canadians did not have paid sick leave as part of their employment, and they were tempted to keep on working even if they felt ill. Only two provinces had embedded guarantees, neither long enough to cover a two-week quarantine: Quebec (two days); Prince Edward Island (one day). Health is a provincial matter, but the provinces claimed that this was a federal problem. Consequently, in September the federal government introduced Canadian Recovery Sickness Benefit (CRSB) that gave $450 a week for up to two weeks leave. But it was criticized for being insufficient and inaccessible since beneficiaries had to apply for it. Angry critics across the country chastised provincial leaders for failing to implement sick pay as a long overdue public-health necessity for COVID-19 and beyond.

Absorbed with the painful, local strife, citizens of Canada, and everywhere else, were only vaguely aware of the relentless expansion of the pandemic to all other corners of the world.

3

Le Tour du Monde:
Latin America, Africa, Asia,
and the Antipodes

By May 2020, cases in Canada were beginning to decline, but Latin America and the Caribbean had become the latest hot spots where the pandemic grew much worse. By early April, the World Bank was providing millions of dollars to support programs for testing and managing the looming pandemic.

Brazil reported its first case on 25 February 2020. Like his American counterpart, President Jair Bolsonaro minimized the danger and did his best to ignore the problem until he himself was infected in July 2020; however, various Brazilian states and municipalities went ahead to impose varying lockdowns and began testing. Without central coordination, the response was chaotic. Numbers of infections mounted rapidly until, by year's end, the country of 209 million posted the second highest caseload in the world, a situation that continued well into the following year. Thanks to its health-care system and its relatively young population, people who fell ill had roughly the same chance of survival as those elsewhere.

Not so for Peru. It had implemented an early nation-wide lockdown, but it had the worst case-mortality rate in the region. Peru has a form of universal health care, but public-health infrastructure is fragmented; hospitals are few and poorly equipped. Furthermore, the aid coming from the World Bank was difficult to distribute since 70 per cent of workers are in the informal economy, and more than 60 per cent of citizens have no bank account, a situation its leaders hope to remedy. Peru's gross domestic product (GDP), with that of Argentina, fell more than 12 per cent in 2020.

Argentina maintained the world's longest, first-wave lockdown. It ran from March to early November 2020. But its health-care system, like that of Peru, is highly fragmented. The lockdown was lifted as the huge caseload finally began to wane with approaching summer. But the privations had delivered severe economic hardship and resulted in more than five hundred reports of police violence, including disappearances and murder, against people accused of infractions. Notwithstanding the lengthy lockdown, by late 2021, following an equally damaging third wave, Argentina had experienced more than 5.8 million cases (almost 12 per cent of its population – a rate equal to Brazil) and 117,000 deaths, the third highest in the region, and it had suffered the worst impact on GDP.

Chile is one of the most prosperous countries in South America with the highest GDP per capita; however, economic disparity is great: sixty thousand millionaires, while 1.4 million Chileans live without toilets or accessible drinking water. The health-care system is a public-private mixture that ostensibly (but unevenly) extends to all citizens. In 2019 and early 2020, riots over the economic hardships coincided with the arrival of COVID-19 in March. The unrest had generated plans for a referendum on constitutional reform, which had to be postponed until October when it won handily. To confront the pandemic, Chile maintained partial, localized lockdowns, curfews, and quarantine. With several isolated Indigenous communities and the remote island of Rapa Nui (Easter Island), outbreaks within the country were scattered and variable. After a sharp, severe, first wave between May and July 2020 and a second wave from December to July 2021, it succeeded in keeping infections and deaths at relatively low rate.

Mexico's first case was reported on 27 February 2020 in a thirty-five-year-old man in Mexico City who had been to Italy; a second case was announced the following day. Soon hospitals were overwhelmed. Health-care professionals began to question the low numbers being released by the government and urged more decisive measures than the mild exhortations to social distancing and hygiene. By early May, the *New York Times* reported that Mexico's case and mortality rates could be three times higher than official statistics. Certainly, Mexico's testing per capita remained the lowest of any OECD nation, meaning that the official case numbers could have been falsely low. Cases cannot be counted if they are not identified with tests, but deaths are harder to ignore. This willful blindness coupled with the disastrous situation in hospitals meant that by early 2021, Mexico's official case fatality rate was the highest in the world at almost 9 per cent: that is, nearly one in every ten diagnosed with COVID-19 would die. This rate was second only to Yemen, a failed state where almost a third of those diagnosed with COVID-19 were dying; that far smaller population is suffering from a decade of conflict, two-thirds of its people are hungry, and clean water supplies are depleted.

As the poorest country in the region, Haiti was one of the first to benefit from World Bank funds. Concerns were high that it would be quickly overwhelmed by the virus. It was still recovering from the devastating 2010 earthquake and a cholera outbreak several months later that had affected eight hundred thousand people and killed nine thousand. Its health-care system is fragile. Haiti's first confirmed case of COVID-19 was reported on 19 March. A tepid recommendation for social distancing and masks was lifted by July, and by the end of 2020, the country of eleven million people had reported ten thousand cases and approximately 250 deaths, one of the lowest fatality rates in the world. Neighbouring Dominican Republic, which hosts many more tourists, had seen almost twenty times the number of cases and ten times the number of deaths. No one fully understands why Haiti's initial figures were so low if they were to be believed. The average age of twenty-two is young, meaning that many cases could be mild or asymptomatic; but the same is true for the Dominican Republic. Could it be that poverty and fear of authority meant that actual case numbers were not properly measured, nor deaths reported? Worse trouble was soon to come in August 2021 when the island suffered another catastrophic earthquake that killed more than two thousand people and destroyed approximately 140,000 buildings or homes. The true state of COVID-19 in Haiti is difficult to know.

Cuba on the other hand, also an island nation with an older average age of around forty years (closer to that of Canada), has an excellent health-care system. It also maintained strict lockdowns that included fines or jail for not wearing masks. By early 2021, it reported fifteen thousand cases and 150 deaths – a case fatality rate of 1 per cent, among the lowest in the region. Hundreds of Cuban doctors went abroad to serve in other countries. But with loosened restrictions in November 2020, cases began to surge, the majority originating in expats visiting from the United States. A far more aggressive wave would begin in mid-2021, bringing its caseload to nearly a million with eight thousand deaths by fall 2021.

AFRICA

After Africa's first case was identified in Egypt on 14 February, observers waited for the continent to explode in a lethal crisis of enormous dimensions. The 2014–16 Ebola fever outbreak in West Africa had been the largest since the disease was first recognized in 1976. Its memory was all too fresh, although many did not realize that traditional methods had contained it. But the African dots on the pandemic maps remained few and small, and the infection seemed slow to develop in this ensemble of developing nations. If anything, it was more prevalent in the wealthier African countries with greater outside contact: South Africa, Egypt. What was happening? Were resources too restrained for testing? Were cases not being counted, investigated, or reported? Were sick people avoiding health care?

At the time of writing, no one can be sure if African figures accurately reflect the prevalence of infection, but they should not be automatically discounted. At first, access to tests was indeed limited, and testing per capita remains low. But the World Bank soon channelled funds into African nations, enabling them to purchase tests and enhance surveillance. Several reasons may account for a less severe COVID-19 experience in Africa.

First, like Haiti, the average age in most African nations is young: ten countries have an average age of sixteen or less; only 3 per cent of the population is over age sixty-five. Second, once retired, elderly people tend to return to their safer rural origins, avoiding crowded cities or care homes. Third, the warm climate is generally favourable for limiting the viral spread. Fourth, the frequent infectious diseases on that continent, especially those of other coronaviruses, may have offered some natural immunity. Fifth, often confronted with outbreaks of infectious disease, Africans have well-developed public-health systems, with good community clinics extending even into isolated areas where they are already well accepted for administering vaccines and treatments. Sixth, the authorities responded quickly with rules that found high levels of compliance in the population: when recommended, masks were adopted with little fuss in most places. Seventh, later in the pandemic, a group reported finding a gene, predominant in peoples of African origin that might make them resist infection. Finally, unlike other warm regions, most African nations did not have floods of tourists, while the few popular destinations lost visitors early in the pandemic owing to border closures and grounded flights.

All this is not to suggest that the pandemic did not present serious problems in Africa. Drug shortages were exacerbated due to market demand, failed supply chains, and ongoing conflicts, not only for medications to treat coronavirus, but also for mainstays in chronic conditions, such as malaria, AIDS, tuberculosis, and mental illness. The latter became a severe problem for months in South Africa, even as the pandemic exacerbated stress.

In Uganda, the musician-politician, Bobi Wine, wrote a song to alert people to the coronavirus in early March; it was quickly picked up by the United Nations Educational, Scientific, and Cultural Organization's (UNESCO) campaign "Don't Go Viral." Nevertheless, the pandemic provided a pretext to further harass the singer, who had become an embattled presidential candidate; he was arrested in November for allegedly violating COVID-19 restrictions at a campaign rally ahead of the 14 January 2021 election, which he lost. In Congo-Brazzaville, another opposition leader, Guy-Brice Parfait Kolélas, died of the virus a day after the March 2021 election, in which he had unsuccessfully challenged President Sassou Ngesso who had held office for thirty-six years with only a five-year hiatus. In the Sahel region, the United Nation's anti-terrorist Operation Barkhane found that the pandemic complicated the mission when insurgents invoked it to terrify local populations. In addition to these issues, COVID-19 poses a huge but as yet undetermined threat in prisons.

A few African countries remained in an unrelenting, low-level first wave. In most, the second wave proved more far more onerous and dangerous.

INDIA

COVID-19 was recognized in Kerala state of southwest India on 27 January 2020 in a twenty-year-old woman who had returned from Wuhan, China, four days earlier. She survived. Sporadic cases soon began appearing. During March–April, the case-fatality rate climbed to 25 per cent, possibly because testing was reserved only for the severely ill or those who had travelled to certain countries. At first, most cases were linked to travel – especially from Dubai or the UK. But a large religious conference, which opened in New Delhi on 13 March, became a superspreader event sending infection throughout the country and to Indonesia, Malaysia, and other lands.

From at least March 2020, India's government and its companies began working on treatments and vaccines seeing it as both a responsibility and an opportunity. Between March and May, India briefly banned export of certain vital drugs, manufactured within its borders, to protect domestic supply. But it too was affected by shortages of protective equipment, especially masks and medications, and by excessive price hikes of finished products and raw materials for drug manufacture from China. During the peak of its first wave, reports of shortages of antivirals were frequent.

But this country, which is rich in pharmaceutical manufacturing, soon had many facilities making and distributing COVID-19 tests, and its public-health authority leapt into action using cluster-containment procedures and denying travel visas. From May, case numbers soared in this the world's second most populous country to reach the second highest load at more than 11.5 million cases by early 2021. But the new cases had peaked in mid-September 2020 and continued to decline during the remainder of the year. Moreover, India remained in the waning phase of its first wave for a long time. Despite its enormous caseload, its per capita rates were all far lower than those of the US or Canada, placing it among the better outcomes globally. This situation generated many epidemiological questions. But all that would change in the spring of 2021 (chapter 9).

OCEANIA

With their proximity and multiple Asian relations in trade and academe, Australia and New Zealand identified coronavirus cases early, on 25 January and 28 February 2020 respectively. Nevertheless, both countries managed the early pandemic with outcomes that were the envy of the world.

Just recovering from a devastating fire season in 2019, Australia has a long history of strict public-health rules and quarantines, as anyone trying to bring a pet to that country will know. Its no-nonsense governments have also long resisted any relaxation on the country's cruel and heavily criticized offshore treatment of refugees. With COVID-19, it quickly clamped down on businesses and schools, and it sealed borders, both externally to the outside world and internally between states and surrounding some vulnerable Indigenous populations. Flights were reduced in terms of numbers and passengers; citizens were stranded abroad, in some cases for more than a year. Its first wave peaked with about five thousand cases daily in late March and a bigger second wave emerged early during its winter in mid-2020, centered on a quarantine hotel in Melbourne. Then cases peaked at about seven hundred daily in late July and declined.

By early 2021 following a second wave in the relatively cooler months of mid-2020, numbers remained relatively stable, schools were open, and it was mostly business as usual, with a total of 130,000 cases and 1,400 deaths. Reactions to any cases at all were sharp and swift. The world was mesmerized by how a single case identified on 7 January 2021 in a cleaner at a Brisbane quarantine hotel, which had housed a cluster, resulted in a snap, three-day lockdown with masking for the entire state. The rigor that Australia applied to contact tracing was on display again when on 13 March 2021, after two weeks with no cases, a Brisbane doctor tested positive after treating two COVID-19 patients in the quarantine hotel. Hospitals, prisons, care homes, and disability centres in the entire region were locked down for a week as authorities followed up on more than four hundred contacts. In late 2021, just as Australia began allowing long-estranged citizens to return, its largest surge began (chapter 12).

New Zealand's experience of the infection was even better. It began imposing restrictions on travel well before it reported a single case. The number of daily cases peaked at ninety in early April 2020, with only sporadic isolated cases thereafter and no deaths at all between 15 September and 15 February 2021. It ended the 2020 year with only 2,340 COVID-19 cases and twenty-six deaths. Like Australia it maintained zero tolerance on the infection. After a blissful run with few cases, the discovery of three infected people in Auckland in late February 2021 resulted in a sharp week-long lockdown of the entire city with no deaths and no spread. But in November 2021 another small surge doubled the number of deaths.

Many reasons explain these good antipodean outcomes. First, they are islands with less cross-border travel and the ability to strictly control travellers. Second, the warm climate is generally favourable to limiting spread. Third, the population is well off and healthy with good medical resources. Finally, the countries acted promptly and thoroughly to invoke public-health measures, and people generally adhered to the rules. Nevertheless, neither country was spared the protests. Demonstrations and rallies against the public-health measures were staged in both countries, mostly led by individuals who perceived the interventions as an infringement on their civil liberties (chapter 11).

These two countries loudly proclaimed an important message: swift, durable implementation of the traditional, non-pharmacological interventions works, and works well. Admired as an example, perhaps, this lesson was one that the rest of the world failed to learn. Table 3.1 illustrates the effect of strict controls. A shorter time from identification of the first case to a peak in case numbers reflects more effective control.

From December 2019 to March 2020, COVID-19 had spread around the entire world. Some countries were able to respond quickly, issuing and following basic rules of hygiene, flattening their curves, and decreasing caseloads in four to six weeks. Elsewhere, the disease smoldered on for up to eight months before showing any decline.

Table 3.1 | Approximate date of first case, date of first-wave peak, and number of days to peak

First case	Country	Peak	Days to peak
31 Dec.	China	15 Feb.	46
16 Jan.	Japan	15 Apr.	90
20 Jan.	US	21 Jul.	183
20 Jan.	South Korea	4 Mar.	44
23 Jan.	Taiwan	20 Mar.	57
24 Jan.	France	17 Apr.	84
25 Jan.	Canada	3 May	99
25 Jan.	Australia	28 Mar.	63
27 Jan.	Germany	4 Apr.	68
30 Jan.	India	16 Sep.	230
31 Jan.	UK	7 Apr.	67
31 Jan.	Spain	1 Apr.	61
31 Jan.	Russia	11 May	101
1 Feb.	Sweden	18 Jun.	138
4 Feb.	Belgium	12 Apr.	68
14 Feb.	Egypt	19 Jun.	126
21 Feb.	Italy	26 Mar.	34
25 Feb.	Brazil	27 Jul.	153
28 Feb.	New Zealand	5 Apr.	37
28 Feb.	Mexico	1 Aug.	155
2 Mar.	Saudi Arabia	19 Jun.	109
3 Mar.	Argentina	22 Oct.	233
5 Mar.	South Africa	19 Jul.	136
10 Mar.	Bolivia	22 Jul.	134

Based on highest seven-day average as reported at CovidNet or JHU, accessed 2 December 2020.

SHORTAGES AND PANIC BUYING

As it ventured from country to country, COVID-19 suddenly put shortages of many kinds on the front page of every newspaper: shortages not only of drugs and inhalers for breathing problems, but also of tests for SARS-COV-2 and especially of personal protective equipment (PPE), such as masks, gloves, and gowns, and cleaning products, including disinfectants. The pandemic unmasked a shocking lack of preparation and provided real-time illustration of the complex interdependency of medical and pharmaceutical supply chains and the dynamics of global actors in competition for scant resources.

Early on, COVID-19 also demonstrated the market effects of panic buying, as people worried about eventual shortages of staples: food, fuel, and personal hygiene supplies, including hand sanitizer, soap, and toilet paper. Nonperishable food items, such as flour, pasta, and tinned goods flew off the shelves, until merchants limited quantities per buyer. This behaviour provoked reassurances from political leaders and manufacturers who pointed out that toilet-paper demand remains the same, pandemic or no, and that the supply in Canada – a land of forests and pulp – was secure. The international obsession with stocking up on toilet paper proved to be a source of mirth in distressing times (figure 3.1).

Grasping individuals took to advertising supplies of masks, hand sanitizer, paper, and Lysol wipes, that they had scooped up in local stores, offering them for resale online at inflated prices. The practice became widespread, and by March, platforms, such as Kijiji and Amazon, had blocked accounts and banned advertising for resold products. From mid-March 2020, most provincial governments provided an online form for citizens to report price gouging. Thousands of complaints were received and many investigated, although it is not clear what consequences were suffered, if any. Most businesses argued that the increased prices reflected scarcities and the added expense for procuring the items. Nevertheless, price hikes were real.

Since masks were widely accepted as essential equipment for healthcare workers, the huge demand caused by managing the pandemic exposed severely limited supplies. Ordinary citizens rushed to purchase them, draining meagre stocks once destined for hospitals, clinics, and testing centres. China had not seemed to suffer these shortages in its first wave – and by late January 2020 in the middle of its own outbreak, it was gearing up to service a massive international demand that grew enormously over the spring. Consequently, China went from making

Figure 3.1 | Editorial cartoon by Theo Moudakis, *Toronto Star*, 11 March 2020 (By permission of the artist.)

twenty million masks daily prepandemic, to 116 million in February, to two hundred million by the end of April 2020. Many Chinese businesses retooled production lines until 76,000 different entities were making masks, the largest of which became BYD Auto.

Europeans were surprised to learn that most of their PPE already came from China, which had been supplying at least half of all such equipment globally prior to the pandemic. With borders closed, air traffic limited, and stiff competition for finite supplies, anxiety turned to outrage in late March, when Europeans learned of American vigilantes who were hijacking the prepurchased stocks bound for Europe on the tarmac of Chinese airports, by offering three and four times the price – in cash.

Food shortages also arose compounding the already dreadful statistics. Before the pandemic one in four people in the world went to bed hungry. During COVID-19, the use of food banks increased everywhere, but donations in kind and in money shrank. Perishable foods were lost when border closures prevented shipments. Sometimes, the problems were to do with inexperience in adjusting supply chains that had been designed to supply restaurants where people were no longer eating. Even in developed nations, prices increased when shutdowns took place in packing plants with outbreaks: for example, the price of chicken in Canada spiked briefly. A study suggested that, in the first two months of the pandemic, the number of Canadians suffering from food insecurity grew by 39 per cent to affect one in seven. The World Bank observed that, over the course of 2020, food prices increased globally (but unevenly) by almost 20 per cent. In July 2020, it also reported that the situation was grievously compounded by the worst African and Middle Eastern infestations of locusts in decades – outbreaks that were driven by climate change.

Shortage of pharmaceuticals was another problem. For at least a decade, the world had already been dealing with a crisis in drug supply. The causes were multiple and stemmed from the nature of the pharmaceutical supply chain, drug prices (both too high and too low), dropout of manufacturers, and temporary changes in the demand-to-supply ratio owing to manufacturing problems, political blockades, and natural disasters. Since 2010 and joined by several colleagues, I had been trying to draw Canada's attention to this problem with an activist website, peer-reviewed articles, and letters to politicians and newspapers, but with little success. The shortages provoked by COVID-19 made the chronic problem seem new and acute.

Without a specific drug treatment for COVID-19, the drugs first deemed to be scarce were those used in ICUs, especially the agents used for sedating patients on ventilators. Temporary shortages of fever-reducing medication and painkillers, such as acetaminophen (also called paracetamol or Tylenol), also appeared. Many people in Europe and America were surprised, and then outraged, to realize that the raw ingredients and many finished products of pharma companies, some owned by their fellow citizens, were actually made in India and China – a situation that had been in place for decades. Xenophobia lurked behind their concerns that foreigners might not honour contracts, or were hoarding supplies, or were letting standards slide. Those coping with chronic diseases – diabetes, epilepsy, arthritis, heart disease, and cancer – worried about the availability of medicines that they needed. In the UK, these fears were exacerbated by impending Brexit. More panic buying, hoarding, and appeals to alternative sources arose. To protect their own supplies some countries (such as France) began taking specific measures to enhance national drug-making capacity; others (India) banned the export of in-demand products that were made at home.

In early April 2020, Canadian pharmacists worried that the drug shortages might increase because of the pandemic's impact on manufacturing and shipping. To conserve stocks, they recommended dispensing smaller amounts of all medications, a long-established practice in dealing with specific shortages. For example, patients who would normally receive a three-month supply would be given a one-month supply and be owed the rest. The policy was met with public anger, magnified by patients wanting to stockpile personal supplies. It was further escalated by the fact that pharmacists would charge the normal dispensing fee each time the client returned, effectively increasing their fees, sometimes by hundreds of dollars for patients relying on several drugs. Pharmacists were intransigent. One-by-one governments leapt in to control the pharmacists' unpopular action, first by covering the dispensing fee for seniors and vulnerable citizens; eventually (perhaps owing to the sticker shock of the charitable gesture), by abolishing the extra fee or the limited dispensing altogether. Pharmacists decried government interference in their professional practice. The plan of limited dispensing had vanished by July 2020. Shortages grew worse and complaints more acute when particular drugs were thought to be good for preventing or treating the disease (chapter 7).

INTERNATIONAL EFFORT BEGINS

At the end of April 2020, an event co-hosted by the director-general of the World Health Organization, the president of France, the president of the European Commission, and the Bill & Melinda Gates Foundation, launched the Access to COVID-19 Tools (ACT) Accelerator. It was to bring together governments, scientists, civil society, businesses, charities, and global health organizations in order to hasten the end of the pandemic. They planned to support the development and equitable distribution of diagnostics, treatments, and eventually, vaccines to reduce mortality and severe disease. They also aimed to facilitate control of COVID-19 and restore full societal and economic activity globally. Recognizing that the virus was highly contagious and more virulent than influenza, mutated variants would constantly cycle back on the countries that had managed to control it. The premise was, "no one is safe until everyone is safe."

Many countries committed financial contributions in the short term, including Canada (4 May) and China (18 May); however, the Trump administration ignored these measures and even withdrew from the WHO in July 2020. The United States government made no contributions to this international effort until after Joe Biden took office in January 2021.

As the next section will show, the COVID-19 response evolved on the legacy of other pandemics, and it marshalled the attention of scientists and policy makers globally to understand and control it.

PART TWO

History and Context

"The history of each disease contributes to the history of all the others."

Mirko Drazen Grmek

4

Cause: Germs, Viruses, Social Determinants, and Ecology

But we are getting ahead of ourselves in this story. People have always encountered disease and tried to understand why it happens and how to manage it. Long before writing began, humans knew that illness could be contagious and the afflicted should be shunned to avoid spread. Even sick animals sometimes slink away. Is it to be alone? To shelter from predators? Or to instinctively protect others of their species? People also sought the causes of disease so that they could prevent it.

In ancient writings, several great epidemics left their traces. In the fifth century BC, Greek historian Thucydides wrote of a plague that ravaged Athens, under siege by Spartan warriors, in *History of the Peloponnesian War*, book two. He described the many theories for its cause, noting that survivors were immune. In the same century, Hippocrates recorded the story of the "Cough of Perinthus" in *Epidemics VI*. Both epidemics have provided juicy fodder for generations of armchair detectives who propose retrospective diagnoses to explain them: influenza, diphtheria, typhoid, smallpox, and anthrax. But in the distant past, those disease concepts did not yet exist.

Even the most famous plague of all – the fourteenth-century pestilence that swept Europe killing millions – had no other name until some six centuries later when a German historian dubbed it the "Black Death."

Aware that illnesses could spread, communities devised methods to prevent and control them. Ancient doctors and military officers understood that sanitation was important and that animals could share human disease. That awareness may have informed the kosher and halal dietary laws. Alexander the Great is said to have kept his troops healthy by moving camp often to avoid contamination of water supplies. Other military leaders located their camps upstream of stables.

The biblical book of Leviticus and early Islamic texts specified how people with contagious diseases must be isolated and avoided. But the first formal public-health measures of quarantine were codified and enacted in fourteenth-century Venice and Ragusa (present-day Dubrovnik), rivals for priority over which city used it first. The word comes from *quaranta* – forty, a number with religious significance, being the number of days of Lent, a period of denial and self-sacrifice. When illness appeared, the possessions and clothing of victims were burned; dwellings fumigated, prayers said (chapter 6).

The familiar notion of germs as a cause of disease also has a long history. In the sixteenth century, the Italian Girolamo Fracastoro wrote about the new European outbreak of syphilis, hypothesizing that the contagious "seeds" of disease must be "alive" – multiplying – to sustain constant spread; otherwise, the infection would disappear.

But no one could see germs with the naked eye or the earliest microscopes. Even after the nineteenth-century technical improvements to lenses, the existence of germs remained controversial. In the late nineteenth century, germ theory – the idea that tiny, living particles provoked human disease – was accepted, following the works of French chemist Louis Pasteur, Scottish surgeon Joseph Lister, and German doctor Robert Koch.* In 1882, Koch proved that a specific bacterium was the cause of a specific disease: tuberculosis. He also laid down four rules, called "postulates," by which observers could establish causal links between other germs and other diseases. We still use those rules.

Koch's four postulates included finding the germ in every case, growing it in pure culture, inoculating it, and recreating the disease in experimental animals. Soon after, many well-recognized diseases with unmistakable symptoms were identified as the products of specific bacteria or other microorganisms. Some germs had been visualized and hypothesized as human pathogens many years earlier; however, Koch's method helped to prove the relationship: dysentery (1883), cholera (1884), typhoid (1884), diphtheria (1884), pneumonia (1886), plague (1894), syphilis (1905). Once germs were known to be the cause of specific diseases, the concept of a "magic bullet" dominated research, as scientists looked for drugs that would kill the germ and leave behind a healthy human. Agents containing mercury, arsenic, and antimony were top of the list (chapter 7).

---

* An asterisk indicates the recipient of a Nobel Prize.

Nevertheless, some very familiar diseases still lacked an identifiable cause, among them, smallpox, chickenpox, measles, influenza, and the common cold. Experiments proved that they could still be transmitted from one animal to another even after the infectious source had passed through a filter that removed bacteria. What could that type of infectious agent be? A chemical? A poison? A toxin? By 1892, the Russian plant biologist Dmitri Ivanovsky chose the Latin word "virus," meaning venom, to describe these unknown agents – whatever they were.

In 1915 and 1917 two researchers – Frederick Twort in England and French-born Félix d'Herelle, who had spent several years in Canada and become a citizen – discovered that in addition to plants, animals, and humans, some of these newly imagined "ultra-microscopic" viruses could also attack bacteria; they were called "bacteriophages." This moment is often cited as the beginning of molecular biology. Their observations spawned research projects into the use of bacteriophages as treatments against bacterial diseases – phage therapy – an agenda that continued until the discovery of penicillin in the 1940s.

Finally, with the 1931 advent of the electron microscope (EM), it became possible to actually see viruses, confirming their existence and wide variability in shape and size. The first virus seen in this way was the tobacco mosaic virus of plants, reported by Ernst Ruska* in 1939. Gradually the viral agents of human diseases were also visualized and distinguished from each other: smallpox (1948), chickenpox (1948), polio (1952). To this day, new viruses are often identified by electron microscopy.

But what were viruses made of? Protein? Carbohydrate (sugars)? Lipids (fats)? Minerals? Were they alive?

By 1939, American scientist Wendell Stanley* had shown that the tobacco mosaic virus was composed of protein and the nucleic acid, RNA (ribonucleic acid). That same year, scientists succeeded in growing viruses in culture on eggs and in animal tissues – a necessary step for studying them and for finding cures and vaccines. To reproduce, viruses take over another organism's metabolism ; they are called "obligate intracellular parasites," meaning that they cannot grow without having other living elements to use. Therefore, viral cultures cannot be truly pure, and that step in Koch's postulates cannot be fulfilled. (Critics have long used this failing to discredit viruses as the cause of some diseases; perhaps the most famous is Peter Duesberg who refused to accept the hypothesis of a virus, HIV [human immunodeficiency virus], as the cause of AIDS.)

In 1944, Nova Scotia-born Oswald Avery and colleagues proved that human chromosomes (and therefore genes) were made of the nucleic acid, DNA (deoxyribonucleic acid). This observation launched a vigorous effort to explain how that molecule could code for inherited factors and how it could replicate in human cells, in bacteria, and in viruses.

From the early twentieth century, X-ray crystallography was developed by the father-and-son team of W.H. and W.L. Bragg* and applied to investigate the ultrastructure of molecules. On this basis, the images taken by Rosalind Franklin in London and obtained (without her knowledge or permission) by James D. Watson* and Francis Crick* led to their famous discovery of the structure of DNA in 1953. By 1955, Franklin had applied her technique to a crystallized tobacco mosaic virus to reveal the world's first full structure of a virus. She published her results in a series of exquisitely detailed and oft-cited articles, that appeared between 1956 and 1959. Tragically she died of ovarian cancer at age thirty-seven in 1958, before receiving any of the plaudits that were her due. In fact, in Watson's bestselling memoir of 1968, *The Double Helix*, she was ridiculed.

With the ability to see, grow, and analyze viruses, the second half of the twentieth century witnessed a virtual explosion of research in virology and molecular biology. Nobel prizes flowed. Viruses were divided into types based on the shape and structure of their protein capsules, the kind of enzymes that they used to harness the host's physiology, and their nucleic acid content, either DNA or RNA.

DNA viruses have either a single or a double strand of that nucleic acid. They include smallpox, herpes (shingles, cold sores, and chickenpox), papillomaviruses (warts and cervical cancer) and various enteroviruses (gastrointestinal disease). RNA viruses include polio, influenza, measles, Hepatitis C, Ebola, HIV, and the coronaviruses.

Coronaviruses represent a huge class of RNA viruses that cause various respiratory ailments, including SARS, MERS, and COVID-19. They were discovered and named by Scottish-born June D. Almeida who moved to Canada in 1956 with her spouse, the Venezuelan-born artist, Henry Almeida. A skilled electron microscopist, she worked at the Ontario Cancer Institute in Toronto studying tumour viruses and publishing on viral structure and classification in the *Canadian Medical Association Journal*. She also developed new staining methods for improving visualization of viruses with University of Toronto immunologists Allan F. Howatson and Bernhard Cinader. Almeida returned to the UK in 1964 and found work at the Common Cold Research Institute in Salisbury. In 1966,

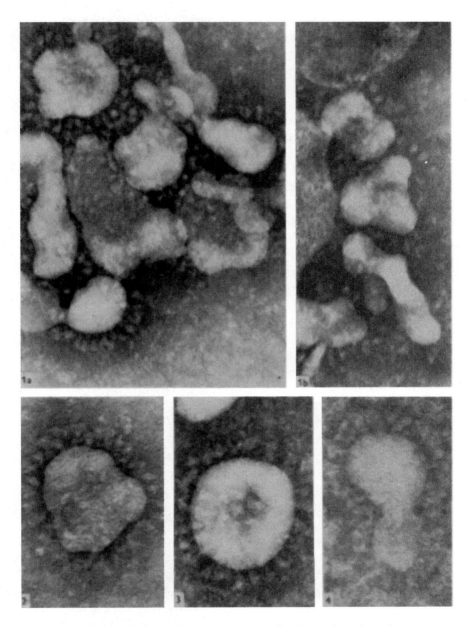

Figure 4.1 | The first images of coronaviruses taken by J. Almeida with an electron microscope (Source: *Journal of General Virology* 1, no. 2 [1967].)

she performed electron microscopy on samples cultured from the nose of a boarding-school boy with an ordinary cold back in 1960. She saw unusual viral particles surrounded by spiky, halo dots (figure 4.1). Convinced that they represented a new family of viruses, related to those affecting birds, she wrote up her results. Skeptical reviewers thought that the blurry images were simply of poor quality; her article was rejected several times before it was published 1967. A year later, Almeida and colleagues named these viruses using the Latin word corona for halo or crown.

DNA and RNA are both made up of chains of four types of chemical building blocks, called nucleotides; their encoded messages depend on the order – or sequence – in which the blocks are arranged. In the 1970s, researchers tried to analyze nucleic acids by identifying their precise composition – a process called gene sequencing. They began with viruses because their chains were relatively short; however, at first the procedure was cumbersome. Nevertheless, a breakthrough came in 1975 out of the laboratory of Bernard Mach in Geneva where the first gene was cloned – a gene coding for rabbit hemoglobin. A flood of gene sequencing soon followed.

The entire chemical makeup of an RNA bacteriophage was published by the Belgian, Walter Fiers, in 1976 and of a DNA bacteriophage by the British chemist, Fred Sanger,* in 1977. Technical advances, discovered and patented by Kary Mullis* in 1983, resulted in the rapid technique of polymerase chain reaction (PCR). Eventually the entire genome sequence of larger organisms could be established: hepatitis C virus (1990), bacteria (1995), fruit flies (1999), humans (2001).

Henceforth, to distinguish between viruses, it mattered less what they looked like – or if they could be seen – and much more if their chemical properties and genetic sequence could be identified. Small mutations in the sequence could make a virus more or less contagious or dangerous (virulent). This mutating characteristic of viruses is why we take flu shots every year.

All RNA viruses are particularly adept at spreading, and they are now considered to be responsible for almost half of the emerging infectious diseases. They are difficult to control because they have short generation times, they mutate far more rapidly than their hosts, and humans share them with wildlife (unlike smallpox virus, which was found only in humans). Coronaviruses, in particular, do not always provoke symptoms in those infected, including animals. They can travel on droplets and on aerosols; some researchers contend that they do not need droplets. The debate over possible airborne spread for COVID-19 was not resolved until 2021 (chapter 6). The dots covering the virus, known as spike proteins, are instrumental in their invasion of cells. Most vaccines aim to attack the spike proteins (chapter 8).

In September 2017, just two years before COVID-19, researchers at the Microbiome Laboratory in Mexico City provided strong evidence for an RNA virus being the agent for "the next global pandemic." As I keep saying, we were warned.

With the arrival of the COVID-19 pandemic, scientists expected that SARS-COV-2 would mutate and develop variants. For the first few months, however, the virus seemed to be stable. But soon the media began to broadcast news of the variants. Notwithstanding the WHO's careful naming of the disease itself and its virus to avoid geographic stigmatization, the variants had the country of origin embedded in their nicknames – UK, South Africa, Brazil, India – until the WHO put a stop to it in late May 2021 (chapter 9). Variants arise where and when the disease is most prevalent because numerous, rapidly multiplying viruses provide opportunities to mutate and spread.

Coronavirus is the necessary cause of COVID-19 – without it the disease does not occur. But alone, the virus is not sufficient – other conditions must prevail to provoke human illness and spread. For example, we still do not really know how old it is, how it crossed the species barrier, and when it started to infect humans. These questions prompt the repeated investigations into the origins of COVID-19 led by WHO and other entities (chapter 12).

Because some viruses infect other mammals as well as humans, an infectious agent will leap from animals to humans under conditions and in locations that favour contact – farms, markets, crowded living arrangements. Influenza is shared with pigs, HIV with monkeys, SARS with bats via civets, and MERS with bats via camels. It appears that the SARS-COV2 virus also came from bats, possibly via pangolins, and it can affect other mammals. In 2019, scientists found evidence that Ebola may also have originated in bats.

Why are bats implicated in these transmissions? Two reasons explain it. First, their highly developed immune systems allow them to carry viruses without being sick; moreover, bat viruses have already adapted to resist immune protections that prevail in humans too. Second, the viruses can spread widely because bats fly. In fact, bats are the only mammals that fly.

SARS-COV2 has been detected in mink, deer, cats, dogs, and zoo animals, but those animals were probably infected by contact with humans. No evidence suggests that they are at the origin of the disease. Nevertheless, in early 2022, Hong Kong found the virus in pet-store hamsters where an employee had tested positive; it ordered a cull of two thousand hamsters – a measure reminiscent of slaughter of pigs in the 2009 swine flu (chapter 1).

Genetics may also influence human susceptibility. Later in the pandemic, research from the Karolinksa Institute suggested that some individuals of African descent possess a gene that offers protection from COVID-19. Still later, scientists in Britain identified a suite of genetic variants that might make things worse.

OTHER HUMANS, HYGIENE, AND THE SOCIAL DETERMINANTS OF HEALTH

As children, everyone learns the social and health importance of washing hands and covering sneezes and coughs. We also know that germs spread when rooms or airplanes are poorly ventilated, when garbage is left uncollected, or when water is contaminated.

Some basic rules of cleanliness and hygiene have a long history, as do the digging and maintenance of sewers and wells. But sanitation, as we know it, originated in the early nineteenth century. Advocates – and not all of them were doctors – maintained that dirty water and filth could cause disease. These arguments gained momentum in the 1840s when William Farr used statistics and maps to explore why diseases, such as smallpox and cholera, appeared in certain places and patterns in England. Also using maps and statistics, John Snow linked an 1854 cholera outbreak to contaminated water from a specific London pump in what was then called Broad Street. From the late 1860s, surgeons were making use of the new concept of antisepsis to prevent infections in surgical operations. Gowns, gloves, masks – PPE – and the gestures of cleanliness, including the ritual of pre-operative handwashing, were codified and practised. These basic sanitarian measures were later consolidated with scientific germ theory in the 1880s, as described above.

A century later, the British epidemiologist Thomas McKeown noted that the declining mortality from tuberculosis in England and Wales had continued downward at a steady rate during the late nineteenth and early twentieth centuries. The advent of new methods for monitoring, control, and treatment, including vaccines and drugs, seemed to have no effect. He argued that the steady rate of decline was owing to an opposite steady rise in the standard of living. Crowding waned. People occupied larger homes and could afford resources to keep themselves and their spaces clean; they possessed their own utensils. The tuberculosis germ was present – and still necessary to provoke disease – but the crowding and poverty that had favoured its spread made the necessary germ less sufficient to cause and spread the disease. Tuberculosis is still with us today, and it causes far too many deaths globally – including in Indigenous communities of Canada. But controlling it requires something more than antibiotics to kill the germ. These additional factors – income, employment, education, literacy, autonomy, nutrition, environment, culture, gender, race, and racism – all play important roles in health inequalities. They are the social determinants of health.

We grew used to reports of COVID-19 total numbers by country or region. But incidence, the rate per capita, was also significant, and its distribution within regions proclaimed the role of these other nonviral causes of disease. When we look at finely grained maps of the distribution of COVID-19, we see that the coronavirus has been helped to spread by the social determinants.

Take, for example, detailed maps of Toronto that express the various community hot spots. At first public-health officials hesitated to release these results, not wanting to stigmatize the most affected neighbourhoods or encourage others elsewhere to lower their guard: the virus was present everywhere; everyone should take care. But the public demanded its release and officials bent to pressure and complied on 26 May 2020.

The map clearly demonstrated that COVID-19 was unevenly distributed in the city. The hardest hit communities were the least well off: poverty was a factor in the spread of disease. People with fewer resources lived with more crowding and worked in situations that forced them to leave home, often using public transit because they did not own cars. After 26 May, the map became interactive – and yet, while the overall rate might go up or down for the city as a whole, the distribution of infection changed little.

Since 2013, the United States had witnessed the rise of the Black Lives Matter movement in America and its spread around the world. Canada too had been wrestling with public admissions of systemic racism in the police, the Royal Canadian Mounted Police (RCMP), and the health-care system. Yet Canadian health researchers had long accepted the notion that race is itself a social, not a biological, construct and that human variation occurs as much within the so-called races as across them. For that reason, the Canadian Institutes for Health Research (CIHI) did not gather race-based statistics, although it did gather information on Indigenous peoples.

The COVID-19 pandemic caused CIHI to reconsider. In spring 2020, it announced new national standards for gathering race-based data to identify inequities and inequalities that stem from racism and its social consequences. Racialized people bear a greater burden of the pandemic, often because their jobs, finances, and living arrangements made them vulnerable. Many were essential workers in long-term care facilities, or in critical infrastructure, such as food industries, grocery retail, transportation, and sanitation. They had to go in person to low-paying jobs to feed families and pay rent. Furthermore, society needed them to keep working. But their work and living conditions placed them at far greater risk of being infected.

Writing in the *Lancet*, Richard Horton urged readers to embrace the concept of "syndemic." First articulated in the 1990s by anthropologist Merrill Singer, the word contains the idea that biological models of infectious diseases would not lead to solutions without parallel consideration of social conditions and other ailments that enhance vulnerabilities. A half century earlier, the physician-historian Henry Sigerist had recognized the same idea, arguing that progress in human health could not occur without balanced improvements in both social and technical medicine. COVID-19 demonstrated the importance of that challenge.

American statistics showed that Blacks and Hispanics had more cases of COVID-19 and were almost three times more likely to die than whites. By mid-January 2021, other studies confirmed that half the population living below median income had generated two-thirds of the deaths. Social factors might matter more than inherited traits of susceptibility or protection.

Politics is yet another social determinant of the spread of COVID-19. In the United States especially, maps revealed that rates of infection and mortality are higher in states with Republican governors (chapter 9). But politics is also a harbinger of preparedness – and lack of preparation allows an outbreak to grow into an epidemic or explode into a pandemic. The political decisions that led to the starving and eventual closing of Canada's Global Public Health Intelligence Network (GPHIN) and vacancy of the chief health surveillance officer led to significant delay in recognizing the emerging threat.

ECOLOGICAL CAUSES OF DISEASE

It is well known that in places where atmospheric pollution is high, the consequences of an infection with any respiratory pathogen is much worse. The same is true for COVID-19. So far, no evidence indicates that

that climate change is a direct cause of COVID-19, although it can easily hamper survival of the infection. Nevertheless, declining biodiversity may play a role in spreading pathogens that we share with animals.

For years, ecologists, like Serge Morand and Kate Jones, have contended that declining biodiversity threatens human health. They argue that deforestation and other land-use practices cause the loss of some species and the continued or enhanced existence of others, harbouring pathogens that also afflict humans. As people move into these newly cleared areas of the rural frontier, the possibility of a leap from animal to human is increased. According to their research, loss of biodiversity correlates strongly with a rise in emerging human diseases of animal origin. Climate change can also diminish biodiversity through habitat loss. The wildlife trade can act in similar ways to disrupt the pathocenosis of our world.

These arguments have been around for a long time, but they gathered momentum following studies of Ebola. With COVID-19, they enjoyed renewed attention. In June 2021, a consortium, through the International Science-Policy Platform on Biodiversity and Ecosystem Services (IPBES), published evidence assembled from hundreds of studies, calling for more holistic methods in clearing land and an end to the wildlife trade. In China, that industry produces animals for food, fur, and medicines and is worth at least US$20 billion annually. China conducted an online survey in February 2020 that showed more than 90 per cent of citizens supported stricter controls on wildlife consumption, trade, and exhibitions; however, some suspect that responses were skewed toward educated urbanites. Nevertheless, on 24 February 2021, China announced a ban on the wildlife trade and compensation for farmers affected by the decision. It is not clear how long the ban will be maintained. A ban on civet meat following SARS was quickly abolished when farmers complained of economic consequences. Others caution that the rules may drive the trade underground, thereby aggravating the risk rather than eliminating it.

If searching for the causes of COVID-19 succeeds in drawing attention to the need to curb the assault on biodiversity, it may also benefit long-frustrated attempts to address climate change.

5

Testing and Tracking

Control of any infectious outbreak depends on the ability to locate hot spots, identify sources and carriers, isolate them quickly, warn contacts, and analyze the effect of measures. That means testing. The tests for COVID-19 came rapidly; their implementation, not so much.

Knowing the cause was not enough. The world also scrambled to understand the characteristics of the new virus. How contagious was it? How long was the incubation period and the period of contagion, during which it could spread? How did it spread? How well could it be controlled by limiting contacts? Note, only the last factor was something that humans could modify. The other factors depended on the virus. In early January 2020, none of this was known about the new disease, although observers of Wuhan and of the cruise ships developed ideas. The uncertainty would wane with experience, but it would not vanish.

Answers to these questions are needed for any pathogen to establish its reproduction number (also called R-nought or R_0). This number reflects duration of infectivity, mode of transmission, and contact rate; it can change during an outbreak. First elaborated in 1952, this arithmetical concept was increasingly applied to infectious-disease modelling by the early 1990s. Mathematical biologists analyze various methods of calculating R_0. If the value is less than one, no epidemic will occur. If R_0 equals one, then without intervention, each infected person would infect just one other, a stable spread would take place; the infection might become endemic, or embedded in the population at a steady rate. An R_0 value greater than one warned of an expanding epidemic. Airborne infections tend to have a higher R_0, because they spread more easily than infections that require direct contact or exchange of fluids, such as Ebola or HIV.

The number can be modified to what is called an "effective reproduction number," or R_t when many members of the population become immune and less susceptible to the pathogen.

By 2019, the reproduction number concept had been applied to the study and control of parasitic infections in both animals and humans and to many diseases of public-health concern: AIDS, measles, tuberculosis, malaria, hepatitis, yellow fever, and respiratory viral infections, including SARS. With the advent of COVID-19, ordinary citizens, previously unaware of the concept, were soon educated on estimates of R_0 for the emerging pandemic. It was difficult to understand how the number could vary for a single virus and differ from one region to another.

The earliest estimates of R_0 for COVID-19 came out of China, placing it at three or more, a high number denoting extreme risk of transmission. Each sick person could infect three others. Experts recommended widespread masking and the invention of rapid methods of testing for the virus to lower the R_0 number and control spread. These reports were submitted for publication in mid-February and published in early March 2020.

THE INVENTION OF TESTS

A diagnosis of COVID-19 cannot now be confirmed without a test. In the beginning, however, no test was available. The disease was recognized as distinct from other acute respiratory ailments by its symptoms and the fact that tests for all other known agents were negative. Infections with the new virus were diagnosed simply because people were sick. The idea that people with no symptoms could be both infect*ed* and infect*ious* was always a possibility. Historians often recall the sad story of Mary Mallon, called "Typhoid Mary," the symptomless cook who unwittingly spread typhoid in food that she prepared for her employers.

Tests for asymptomatic disease became routine in the twentieth century for work, school, and travel. They were used to screen people for tuberculosis and syphilis, and to check donated blood for hepatitis, HIV, and other infections. At the beginning of COVID-19, no one could tell how many apparently healthy people might have this new virus and for how long they might be infectious.

Only with the sequencing of the genome for the virus were scientists able to create tests. That sequencing came quickly, and Chinese researchers released it to the world by 10 January 2020. Consequently, scientists in many different countries were able to start making diagnostic tests. It quickly emerged that many people could be infected with the virus and yet remain symptom-free; often they were young, healthy, active. The experience of the *Diamond Princess* cruise ship also hinted at the high proportion of hidden infections in an older population (chapter 1). To control this pandemic, the world needed reliable tests. Millions of tests. Quickly.

Within two weeks of the sequencing, various tests were created in China, Malaysia, Hong Kong, England, Germany, Russia, South Korea, and the United States. Ontario made and used a test as early as 12 January 2020 on someone who returned from China with symptoms – it was negative. Was that a true result? Or was this precocious test simply not reliable?

Having created a test, logistical problems ensued. Facilities for the rapid production of massive quantities of test kits had to be built. Distribution to consumers was complicated, given travel restrictions and border closures. Samples from noses and throats had to be collected by trained workers wearing protective garb to avoid becoming infected themselves. By late February 2020, South Korea quickly implemented massive testing with a drive-through option to enhance comfort and safety. The gathered samples then needed to be read or interpreted by experts in a laboratory, and the results would have to be communicated in a reliable, timely fashion to be any use at all. A negative report from a sample seven days old means very little in an active outbreak.

The tests invented for COVID-19 were based on methods that had long been established. They were basically of four types. The first type is a molecular test, called PCR (for polymerase chain reaction), based on the sequencing techniques discovered by Mullis in 1983 (chapter 4). The PCR test is designed to detect the active presence of the coronavirus's RNA in the patient's secretions, usually performed on a nasal or throat swab.

The second is an antigen test, which uses monoclonal antibodies to detect proteins on the surface of the virus. Less reliable but faster and cheaper than the other tests, the antigen tests grew in importance later in the pandemic.

The third, is an antibody test, performed on the patient's blood, designed to detect an immune response that the patient might have made against an earlier exposure to the virus. Antibody tests do not control active infections; they are used by clinicians who are concerned about patients' immune defences, and by researchers who want to determine the strength and duration of immunity from infection or vaccines.

Finally, a far more complex and time-consuming test called genetic sequencing can be performed on actual viruses isolated from patients. It is designed to assess the genetic makeup of the virus, to search for and identify variants. Most centres used genetic tests on a small percentage of infected samples to track the spread of variants.

No test is perfect. All diagnostic tests in medicine are assessed for the virtues of *sensitivity* and *specificity*, and they have different thresholds. A *sensitive* test will pick up every case, even if the virus or antibody levels are low; it has few false negatives, but it might have false positives. A *specific* test will pick up only true cases; it will have few false positives, but it might have false negatives. A manufacturer will work to ensure that a test is both sensitive and specific. But inevitably, some products are better than others. One way to avoid problems is to use a sensitive test for screening, and a specific test to confirm – a system of two tests that had been implemented early in the control of AIDS. In a rapidly moving pandemic with a new pathogen, caregivers can only dream of such luxury.

In addition, the chance of finding a positive test depends not only on the type of test, but also on the timing of an individual's sample with respect to the moment of exposure or of developing symptoms. The number of virus particles after infection rises for about a week, reaches a peak around the time symptoms begin, and then slowly falls off. Antibodies that we make respond to the virus and make us immune. Various types of antibodies follow different schedules: first come the larger IgM antibodies found mostly in blood, followed by the smaller, more ubiquitous IgG antibodies that last longer. Consequently, a PCR test (for the virus itself) has a higher chance of being positive at the end of the first week of the infection when the viral load is highest. Antigen tests are most reliable when the viral load is high in people who have symptoms. An antibody test will not likely be positive until a bit later and the positivity may persist for some time long after the infection is contagious. In other words, the sensitivity and specificity of even excellent tests vary throughout the course of an illness.

One pandemic uncertainty, quickly clarified, was that infected people could spread the virus two days before their symptoms appeared. Hence, testing after an exposure was important; however, the viral load might be too small to be detected yet still be contagious. A later test was usually recommended if not required for those who had been in contact.

By February 2020, commercial laboratories were making and selling test kits for the novel coronavirus. In just five days, the giant Chinese genomic company, BGI, had erected a massive testing centre for Wuhan;

other sites quickly appeared around the country and sold products to other countries.

During February 2020, the United States Centers for Disease Control (CDC) manufactured test kits and blocked other laboratories from offering tests hoping to ensure reliability. But only about one hundred CDC tests could be processed each day; by the end of the month, just a few dozen cases had been identified, almost all in travellers with symptoms. But the CDC soon learned that some of its tests were faulty and had to be recalled; it stopped shipment of its tests to other countries. By March 2020, the CDC allowed other labs to use its kits, and commercial entities entered the game as well. The Food and Drug Administration (FDA) began issuing Emergency Use Licenses (EULs) to at least six companies for their PCR-test products.

There weren't enough tests. Strategic decisions had to be taken to stretch supplies. Some jurisdictions decided to test only those people who had symptoms. Some were testing returning travellers or close contacts of those who had tested positive.

Once the suspicion of asymptomatic infection, described above, was widely accepted – in early March 2020 – the idea of testing only symptomatic people became flagrantly inadequate. By October 2020, molecular physiologist, Michael Kotlikoff, provost of Cornell University, argued that infection without symptoms might be as high as 80 per cent in the healthy student population. He contended that the CDC's original suppression of testing by well-prepared academic institutions had cost many lives.

Having declared a pandemic and recognized asymptomatic spread, the WHO called for member nations to ramp up testing capacity, further escalating global demand in the face of short supply. In the late Canadian winter, people who were ill had to wait hours, or even days, to obtain a test, sometimes standing outside in long, cold lines. Many were too sick to bother. Gradually more facilities became available, and some countries set out to test as many people as possible, hoping to define and track the extent of COVID-19 within their borders.

Being able to display a negative test result became essential for children with stuffy noses to attend school or for workers to go back to work. Eventually negative tests were also required for boarding airplanes and entering countries. Thousands of workers in full PPE were engaged in the enterprise of testing: greeting contacts, directing traffic flow, carefully recording health data, swabbing, labelling, shipping, analyzing, and communicating the results. These workers were at risk all day, every day, but many remarked that they remained cheerful, patient, and sympathetic. At the test centre in Kingston, Ontario, a hand-made poster stuck on the wall stated the obvious: not all heroes wear capes.

The PCR test was the gold standard and the earliest available, but it requires a swab taken from the back of the nose. It is uncomfortable, and squirming, frightened children found it painful. Throughout the pandemic many simpler and faster tests were devised based on samples swabbed from the throat, saliva, or lower in the nose. By a year into the pandemic, various kits were commercialized for personalized use around the world. A team in Vancouver announced a test based on breath. Promising results within minutes, the newer tests were useful for people working or living in congregate settings where repeat testing became necessary.

By late 2020, various companies had prepared rapid antigen kits designed to be used at home and to facilitate the need to document negativity for work or travel. These products needed approval from federal and regional authorities – approvals that often took much longer than designing the test. But since they were antigen tests, hopefully with reasonably high sensitivity and specificity, a negative test might be somewhat reassuring, but positive results would have to be confirmed with a PCR test.

Because of their lesser accuracy, the rapid antigen tests were relatively unpopular both with scientists and the public. Later in the pandemic, however, with the arrival of the highly contagious *omicron* variant, the official PCR testing sites were overwhelmed. Demand for the do-it-yourself rapid antigen tests soared; they were better than nothing, especially for people with symptoms. Once again, however, shortages emerged (chapter 12).

TRACKING THE SPREAD

The advent of tests for the new virus, coupled with the power of computers and the internet, allowed the COVID-19 pandemic to be tracked by everyone in real time. But how was the new disease followed internationally? What determined the March 2020 decision to finally label it a pandemic? And could how we have confidence in what we were being told?

Consumers of media soon became inured to seeing a real-time, interactive map of the world covered in dots of varying size that reflected the number of known infections in each country. It was mesmerizing and horrifying, like watching a multivehicle pileup in slow motion, as the dots expanded in number and size, spreading from continent to continent.

The WHO operated one such dashboard, as it already did for many other diseases. Agencies in individual countries, the American CDC and Health Canada, did the same for their own regions. But perhaps the dashboard most often cited in the media, and therefore the most familiar, was that built by engineer Lauren Gardner, a professor in the Center for Systems Science and Engineering at Johns Hopkins University (JHU).

First shared publicly on 22 January 2020, the JHU tracker showed a world map in black swith red dots and ancillary charts to portray actual and logarithmic numbers of mounting cases and deaths through time for each country and the entire world. It also reported data at the state and province level for the United States and Canada. Professors at my university were using it in their information sessions before the end of the month. By 19 February (in print on 1 May), the owners boasted in the *Lancet* that its data compared favourably with sites of the WHO or the Chinese Center for Disease Control. They committed to making it freely available for the entire duration of the pandemic.

Another excellent tracking site emerged, created by a group of first-generation Chinese immigrants to the United States, led by Harvard-educated biostatistician, Dr Yu Guo, originally from Wuhan. She had lost family members to the disease and wanted to help. Her team built CovidNet "to bring more transparency to the public and increase awareness about the global epidemic." Initially focusing on the United States and Canada, they eventually covered global news and many other countries. Their website included interactive graphs that displayed the comparative rise and fall of national (or state) case numbers from day zero to the present. Their data has been used by the CDC, WHO, and academics, and they published technical reports on their operations. The high traffic at the open-access site, with no funding, led them to resort to crowd sourcing through a pop up window: "Buy us a boba tea."

Other entities, such as the social enterprise BlueDot founded in 2013 by University of Toronto infectious disease expert, Dr Kamran Khan, offer free information and subscriptions to governments and businesses to track and manage the pandemic response. From 2003, Khan had spent a decade creating a blend of geographic information, natural language processing, and artificial intelligence to build early warning systems capable of predicting outbreaks. Honoured with one of the Governor

General's Innovation Awards in 2018, Khan saw his BlueDot receive more investments and staff during COVID-19. Who knew that there could be such hunger for information and competition in this domain?

At first, the variation between the figures in these various dashboards seemed significant. The differences reflected the methods of detection, the number and accuracy of the sources, and the speed at which information was conveyed to the provider and transmitted to the web. The JHU site is updated several times a day; the others daily. Weekend counts are lower; Monday reports can be high to compensate. The seven-day average of new case counts more reliably reflects trends. The dashboard with the earliest report of an affected region and the highest case count seemed to be the best. But could we trust how the cases been identified? Were they presumed only, or were they confirmed by testing? If by testing, how reliable was it? Meanwhile the contagion was an ever-expanding target.

As the pandemic grew, the differences in the dashboards became negligible. For example, on a November 2020 afternoon, the global numbers at three sites (WHO, JHU, CovidNet) differed by a half million cases and six thousand deaths. Minutes later, the JHU site case count increased by more than 150,000; the deaths by more than two thousand. The two other sites caught up the next day. I wondered how much higher the numbers would rise before the pandemic would end. I have to admit that at that time I had no idea that by June 2022, the global figures would rise to half a billion cases and more than six million deaths. With numbers on that scale, originally large differences between tracking sites vanished completely.

The data-gathering sites provided other comparative information beyond daily and cumulative numbers of cases and deaths. They allowed for a per capita analysis, or incidence rate, to compare the extent of disease from region to region, and for a case-fatality rate showing what proportion of confirmed cases had died (chapter 9). The latter reflected how well the health-care systems of various localities were coping with their sick – a measure of the likelihood of surviving the infection.

But the case fatality rate is meaningful only where testing is abundant and accurate, and authorities are willing to be transparent. People did not stop dying of other diseases and accidents. For this reason, epidemiologists would examine excess deaths over usual predictions for the same place at the same time of year. One way to allay fears and bolster confidence in the local ability to manage the disease was to downplay the national or regional death rate by using narrow criteria to identify which deaths were caused by COVID-19. Politics again played a role, as comparisons between Russia and Belgium would show (chapter 9 and table 9.1). Avoiding narrow definitions, simple numbers of excess deaths could tell an entirely different story. Two years into the pandemic, several agencies, including the WHO, concluded that excess deaths worldwide were up to three times greater than the official COVID-19 statistics. The number would include undiagnosed and unmeasured cases of COVID-19, but also some deaths from other diseases provoked by the pandemic strain on hospitals, medical supplies, and finances.

While the disease surged globally, China's epidemic peaked, declined, and simmered among the lowest totals and incidence rates in the world. Nevertheless, the case-fatality rate seen in Hunan province in early 2020 would stand as one of the highest at about 6.6 per cent. Most cases identified in that province had appeared before any testing or special treatment was available. Asymptomatic cases and probably many others with mild symptoms were simply not counted because they could not be identified. Therefore, that case-fatality rate is deceptively high.

Another element of testing entailed the positivity rate. If a large percentage of the tests were producing positive results, observers could be certain that many infections were not being diagnosed and could lead to more spread. Ideally the volume of testing would be considered adequate if 1 per cent or less of tests were positive. In surges, the percentage would rise and sometimes testing simply could not keep up.

Being able to test and identify cases is one thing; telling other nations, though equally important, is often viewed as something entirely different. Unwillingness to reveal rates of infection or death had posed a thorny problem in the nineteenth-century Sanitary Conferences (chapter 1). Openness demands mutual trust and respect, commodities all too often in short supply – even now. To protect trade and the economy, a country might not be willing to admit to an epidemic. It might deny or falsely diminish a death rate to bolster confidence in national security and health-care provision, both at home and abroad. The degree of transparency varies between countries and sometimes within them, between provinces and states. These political concerns often placed frontline workers and public-health experts in direct conflict with government authorities.

When reports proclaimed that a country had few infections or a low death rate, confidence in the good news would be raised by a high level of testing. But as happened in Wuhan, and also in Russia, doctors who expressed concern over the official statistics or the disease management could face arrest, beatings, and death. False or incomplete reports stood unchallenged.

Two years into the pandemic, tests continued to be vital for COVID-19 control, but problems persisted with their availability, use, and cost. Creative solutions were found for testing entire communities, rather than individuals, by applying PCR tests to wastewater. Because the virus is excreted in feces, it can be detected in sewage and monitored to predict where and when outbreaks might occur (chapter 12). Like the familiar ultraviolet (UV) index or air quality index, these public-health warnings could be used for collectivities at risk.

6

Controls and Their Side Effects

The initial response to the pandemic was traditional. Public-health authorities reminded everyone to use the basic rules of hygiene: wash or sanitize hands frequently; cover coughs and sneezes with an elbow, not a hand; do not touch face or eyes; keep physical distance from others; and avoid touching, hugs, kisses, handshakes, and large gatherings. These rules extended to specifics about the exact distance to be maintained from others whenever possible: ideally two metres. For those who felt ill or were deemed infected by positive tests, isolation and quarantine were implemented to prevent spread.

The simple gesture of handwashing occasioned instructions on how to do it properly. In early March 2020, Prime Minister Boris Johnson urged fellow Britons to wash hands vigorously while singing "Happy Birthday" twice. The handwashing obsession sparked a CBC Radio *Ideas* program on its history (28 May 2020). Hand sanitizer became a respected alternative, and a boom in sales prompted shortages. But it soon emerged that not all hand sanitizers were created equal. Fraudulent versions kept investigators busy throughout the year. Some of us wondered if even more reliable hand sanitizers could backfire by providing a false sense of security that resulted in laxity with other essential precautions. We recalled how other supposedly antiseptic fluids of the past had inadvertently spread germs or harmed human physiology and the environment; consequently, the US FDA had banned antibacterial soaps in 2016. Killing bacteria is one thing; killing viruses has always been much more difficult. A high concentration of alcohol is key; it weakens the outer capsule of viruses and reduces their virulence. Consumers were instructed to read the fine print on the hand sanitizer labels.

Scientists began reporting on how far droplets could travel in talking, sneezing, coughing, singing, and playing musical instruments. They studied videos of air leaks around face masks and shields. They analyzed how long aerosolized particles could hang about suspended in the air, and they calculated their rate of turnover in spaces of various ventilations. Disinfecting all objects and surfaces likely to have been in contact with the virus was also deemed important, and trials were established to determine if the virus could remain contagious on inanimate surfaces, such as counter tops, sinks, and toilets. Like other coronaviruses, SARS-COV-2 was excreted in feces. Extra care must be taken with bathroom hygiene and ventilation: people were reminded to close the lid before flushing.

Fresh and circulating air was important, as were filters for enclosed spaces, such as airplanes and cruise ships. Qingyan Chen, an engineering professor at Purdue University, had already studied the transmission of SARS in airplanes and cruise ships. He noted that most airplane filters would exclude viruses, but cruise ships filters were inadequate, even under lockdown. If an infected passenger flushed without closing the lid, the virus could easily enter the air-conditioning system and spread from cabin to cabin. He was interviewed dozens of times on this matter during the *Diamond Princess* outbreak and after.

Airlines and airports implemented strict controls for temperature checks, masks, and sanitizing hands and equipment. Studies were published on the ventilation of aircraft, which mostly showed good turnover. Occasional examples of in-flight contamination were scrutinized for how the virus propagated between seated passengers. Later, proof of a recent negative test became obligatory before boarding a plane, and some airports introduced rapid testing on all arriving passengers. Nevertheless, the virus could still escape detection.

How COVID-19 spread through airborne transmission continued to be debated well into 2021. The discussion hinged on the differences between droplets, which are larger and contain most of the exhaled virus, and aerosols, which are tiny and hold much smaller amounts of virus. In normal conversation, droplets usually fall to the ground within the famous two-metre distance. But aerosols can persist for a long time, floating and accumulating in enclosed spaces, especially those with poor ventilation.

An argument developed in the journal, *Clinical Infectious Disease*, during the summer of 2020. It was waged between an international group of scientists who were interested in the physics of air flow and aerosols, led by physicist Lidia Morawska of Brisbane, Australia, and a group of infectious disease clinicians in Toronto, Canada. The former group contended that airborne transmission was a distinct possibility, which was being ignored simply because clinicians wished to avoid panic; these physicists invoked the names of 236 scientists who agreed with them, including Quinyan Chen. The clinician group argued that it could not support the view for lack of evidence and the fact that vigorous application of the rules of hygiene, based on the droplet theory, was already shown to be effective in bending and flattening the curve. The spat was a debate pitting scientific purists against pragmatic clinicians who were toiling in the COVID trenches and who resented the insulting implication that they were ignoring science. At the time, Brisbane had relatively few cases of COVID-19; Toronto was in an emergency state, but its measures were working.

At first, face masks were controversial, especially in Europe and the Americas, but not in Asia. For many years, people in China, Japan, the Philippines, and Taiwan adopted face masks as a matter of course, protecting the wearer from atmospheric pollutants and germs in crowded conditions. Japanese people with colds wore masks to protect others; it had become an element of good manners. Some women wore them if they had not applied makeup. From the earliest days of COVID-19, compliance in Asia with recommendations for wearing masks was impressively high. In fact, people did it without being told.

Beyond those countries, masks had been seen only rarely in open society. Professionals wore them in hospital isolation units, dental offices, operating rooms, and dusty workshops. Sometimes, my patients wore masks in public if their immune systems were suppressed by cancer chemotherapy or AIDS. But those vulnerable people usually shunned the out-of-doors and put on their masks only as they scooted between clinic and home.

From early spring 2020, whether or not masks should be adopted by everyone became a contentious issue, scientifically, politically, and also culturally in Europe and North America. At first, when most cases outside China were identified in returning travellers, scientists doubted their utility because they did not yet have evidence of airborne, community spread – and few cases were around. The experts did not actually oppose the wearing of masks, but they could not promise that masks would be doing anything at all for the wearer or for others when no virus was circulating. Shortages of masks meant that the scant resources should be spared for health-care workers. Some public-health experts also worried that people might think that donning a mask, like using hand sanitizer, allowed them to relax the other gestures of hygiene and distancing that were also known to be effective.

Later, with strong evidence of community spread and the spectre of asymptomatic carriers, public-health officers recommended masks. By then, the reason was not only to protect the wearer of the mask but to protect everyone else from the wearer who could easily have the virus and not know it.

This flip-flop in recommendations became a rallying cry for those who opposed any controls at all. It was cited as evidence of inconsistency, ignorance, and intrusive interference in private lives. It was used to mock and discount all other scientific recommendations: since they were wrong on masks, they're wrong on everything else. Furthermore, masks were for wimps and weaklings. Masks proclaimed the wearers' fear of infection, their subservience to suspect rules, and their belief in a threat that had been discounted as fake: even the president of the United States insisted that COVID-19 was nothing but a cold. Not wearing a mask became a political statement, a trope of civil liberty and strength, terrifyingly displayed by the US president throughout his lengthy electoral campaign and especially on 5 October 2020 when he stood, infected, on the White House balcony and dramatically demasked himself in cavalier disregard for the journalists below (figure 6.1).

Other currents in Canada added extra fuel to this fire. In early April 2020, when my daughter and son-in-law donned the by-now-recommended masks to enter an Ontario pharmacy to pick up medicine for their child, a man going out the door expressed his disgust: "Fucking masks!" he muttered at them.

Quebec, in particular, had enacted legislation against all face coverings in 2017. It was soon overturned by arguments that the law had unfairly targeted Muslim women. To replace the law in June 2019, less than a year before the pandemic, the new provincial government passed *An Act respecting the laicity of the State*. The controversial legislation banned the wearing of religious symbols in public; the section on face coverings was under debate and left suspended. Many people criticized the new law for its racism and xenophobia. For those who supported it, a recommendation to don a mask for health reasons seemed absurd, even antinationalist. For those who opposed the legislation, the possibility that masks might now save lives contained a delicious irony that only deepened when Quebec made mask wearing mandatory in July 2020.

Figure 6.1 | Editorial cartoon by Kevin Necessary, 19 April 2020
(With special thanks to the artist.)

This problem – convincing unwilling citizens to behave in their own best interests – is a hazard of public-health work. It relates closely to the opposition that the WHO has always encountered on a much grander scale (chapters 1 and 2). Confronted with a completely new disease that was caused by a completely new pathogen, new information appeared daily: it was an ever-changing scenario and recommendations could only follow the available evidence. As the evidence changed, so should the rules. Scientists needed to be vigilant, nimble, versatile, and convincing. Constant derision and pushback over changing regulations kept reminding me of the famous quip attributed to J.M. Keynes: "When the facts change, I change my mind. What do you do, sir?"

QUARANTINE AND ISOLATION

Originating in the medieval plagues of Europe, quarantine is a wait-and-see manoeuvre. It isolates the infected – or potentially infected – from the rest of the population. In its first instances, at the ports of Ragusa (now Dubrovnik) and Venice, ships were to drop anchor offshore and wait forty days before docking and unloading (chapter 4). Should the disease be present on the ship, it ought to have burned itself out within those forty days. Unfortunately, the people of those times did not understand the role of vectors – rats, fleas, and germs – in the spread of bubonic plague. The infection often escaped anyway, riding on the rafts sent back and forth to provision those confined on board. Various forms of quarantine were widely applied in other ports and inland cities. Its draconian adaptations within Milan's closed city gates probably accounted for that city's relatively fewer deaths in fourteenth-century plague.

Quarantine has been used regularly over the centuries and into our time to prevent the spread of illness. As travellers with pets to Australia and England know well, it is implemented for up to six months to protect those island realms from rabies – successfully so far. But the precise methods of quarantine also rely on scientific information. When used as a blunt instrument, they sometimes lead to inadvertent spread of contagion or unjust infringements on minorities, providing more lessons to inform the next crisis.

In the mid-nineteenth century, Canada developed a quarantine station on Grosse Île in the St Lawrence River off Quebec City. Immigrants arriving from Europe were known to be possible carriers of cholera and typhus. The passengers were taken to the island and housed in barracks, sorted by the economic class of their tickets, rather than by the presence (or not) of illness. Crowded in these makeshift warehouses of varying luxury, those who had not been ill at disembarkation soon sickened, and many died. The Grosse Île National Historic Site marks the mass grave of at least six thousand people. Similar errors of failing to separate sick travellers from the well were made at many nineteenth-century ports.

Contact tracing is another well-established method to control the spread of infectious disease. It had been used in Milan to control plague in the fifteenth century, and again in Marseille in the eighteenth century. Authorities understood that contacts should be kept separate from the sick and the rest of society, although that ideal was honoured more in the breach than in practice. They would try to identify anyone who had been in close contact with someone who was sick. Only in the nineteenth century was there a greater awareness – and accompanying sense of helplessness – about the idea that people could be infected with disease (such as influenza or typhoid) and display few symptoms or none at all.

Contact tracing was used extensively throughout the WHO's successful campaign to eradicate smallpox, and it has also been used in leprosy, tuberculosis, and Ebola fever control. It grew familiar during and after World War II for identifying the sexual partners of those with venereal diseases – syphilis, gonorrhea, and later HIV and AIDS. In that capacity, it became somewhat notorious, since it required intrusion into the intimate details of private lives. I vividly recall being sent as a medical student to observe the early 1970s venereal disease clinics of Toronto, run by well-intentioned and sometimes zealous staff, where unhappy infected people sat, tortured and ashamed, unable to utter the names of their contacts out of embarrassment, or because they had never known the names in the first place.

For COVID-19, the aim of quarantine was to *isolate* those known to be sick or testing positive, from those who were healthy and tested negative. Everyone who had been in contact with a known case – or who had travelled into a closed jurisdiction – was to quarantine for 14 days and hopefully be tested. Later, in some places, the wait was reduced to ten, seven, or five days. Still later with more information and the threat of variant strains of virus, the wait was increased once again. When and where tests were readily available, a negative test, taken at seven or fourteen days, or both, after exposure, would permit an end to the confinement. Every jurisdiction set different rules. Lacking specialized centres for quarantining massive numbers of healthy contacts, authorities allowed and trusted people to quarantine at home. When this proved impossible, owing to living arrangements or lax observations, hotels were used, sometimes at the contacts' own expense. For returning travellers, the rules were the same as for contacts. Stiff penalties were legislated for those refusing to respect the rules: in September 2020, Ontario announced fines of up to C$5,000 a day for contacts who refused to isolate. As the pandemic rolled on, fines were adjusted upwards to reach $100,000 maximum or a year in prison for individuals who defied orders.

Contact tracing became sophisticated, and many countries made use of cell-phone apps. Citizens could choose to place an app on their devices that would alert them anonymously if they had been in proximity with an (unidentified) person who had tested positive. The method was entirely voluntary, but it relied on a certain willingness of citizens to allow this intrusion into their privacy. Several countries in Europe and North America adopted it, recognizing both its limits and benefits. Naturally, many viewed the apps for digital surveillance as a form of spying and refused to contaminate their phones. In Quebec, opposition parties protested the plans, arguing successfully that such an app would result in far too many false positives, provoke unnecessary panic, and lay an added burden on the testing system because it could not tell whether cases and contacts had been wearing masks. In early August 2021, a group of Italian researchers, led by economist Margherita Russo, published (in *Vox EU*) an in-depth comparison of various apps across eighteen countries, the EU, and the WHO; they could find no evidence that the technology had had any effect on controlling COVID-19. The apps were abandoned.

Using both digital and traditional in-person work, several countries became exemplars in contact tracing with excellent results at least during 2020: China, South Korea, Taiwan, Bhutan, Vietnam, and Japan (table 9.1). The first WHO report described China's "meticulous" system in Wuhan "where more than 1,800 teams of epidemiologists, with a minimum of five people per team, [traced] tens of thousands of contacts a day." These countries also provided financial support and shelter for those in quarantine. South Korea not only traced contacts forward in time from positive cases, it also tried diligently to trace contacts backwards from each positive case to locate the person(s) who had been the source of the infection and to find their contacts too. With the advent of the more aggressive *delta* variant in 2021 and the economic hardship wrought by lockdowns and limited trade, Vietnam's situation deteriorated to one of the worst in the region despite its brilliant start.

But wealthier countries, with their long underfunded public-health systems, had trouble marshalling the human resources needed to do contact tracing properly and the social will to comply. By November 2020, *Nature* estimated that more than half the positive cases in the United States refused to name their contacts, and a large proportion of those identified were never reached by the tracers. In the UK, tracers failed to reach more than 10 per cent of positive cases and more than half of their contacts, while nearly a fifth of cases refused to divulge information about their contacts. Well over half of those ordered to quarantine admitted that they had been leaving their homes.

In the situation of an overwhelming surge in cases, many jurisdictions simply abandoned contact tracing to focus on controlling and caring for the positive cases – meaning that the virus could spread at will: for example, various states in India (July 2020); Slovenia (October 2020); the United States (South Carolina, January 2021); and in Canada: Toronto (October 2020), Alberta (November 2020), Saskatchewan (September 2021), Ontario and Quebec (December 2021). In the absence of adequate testing and contact tracing, officials relied on numbers of hospitalizations to track the extent and severity of the pandemic.

Rare were the regions that turned to volunteer contact tracers. The idea met with objections from professionals who feared that unpaid workers would lead to some future denigration of the value of their skills. When things heated up in Ontario in the late fall of 2020, an appeal went out from Kingston, Frontenac, Lennox, and Addington (KFLA) Public Health for volunteer contact tracers. I stepped up along with about two hundred others; the fifteen selected were all retired health-care professionals – nurses, doctors, pharmacists. Together we experienced the complex and surprisingly detailed training needed to approach the task and the even more complicated relationship with the database platform for epidemiological reporting. The phone calls were the easy part – often poignant, sometimes tense, although most people understood and accepted the need to isolate. Even if it meant loss of income, they complied gracefully without being reminded of the $5,000 daily fine. Through several more waves, we also helped other regions with surges, and the professional nurses expressed gratitude. KFLA carried on contact tracing until late 2021 when the rising numbers and inadequate testing made it impossible (chapter 12).

LOCKDOWNS: THEY WORK

A lockdown means different things in different jurisdictions. It might entail some (or all) of the following measures: stay-at-home and work-from-home orders, closure of external and internal borders, closure of schools, of all nonessential businesses and manufacturing, limits on distance travelled from home, limits on public transit, cancelling of mass gatherings for sport, entertainment, or religious reasons. They are coupled with rules for isolating cases, quarantine for contacts and travellers, and various curfews on evenings and weekends.

Lockdowns – national or regional – were issued in most countries confronted with COVID-19. Wuhan and Hubei province were the earliest and tightest, but as Europe became infected, countries implemented the orders one by one following the march of the disease: Italy (16 February municipal, 8 March regional); Poland (10 March), Spain (14 March); France (16 March); United Kingdom (23 March). Germany dealt with the situation regionally with partial or complete closures of schools and borders. The Czech Republic took a middling approach installing curfews and border closures. Beginning with a duration of two weeks, these lockdowns saw repeated extensions. Argentina issued the earliest lockdown in South America (on 19 March); it eventually became the longest in the world.

Some countries, such as Sweden and the United Kingdom (UK), adopted a more laissez-faire attitude believing that, because universal infection was inevitable and survival rate seemed to be tolerably high, allowing the infection to simply run its course would be the best plan. They reinforced the rules of hygiene and social distancing but were variably permissive for other activities. They did not absorb the logic of flattening the curve. Although the leaders later denied it, they touted the virtue of herd immunity, a condition arrived at when 70 per cent or more of the population is immune. It is usually achieved only with mass vaccination, not mass illness. The later denial of this policy fed the justified anger of Richard Horton, editor of the *Lancet,* who decried it as a form of state-sponsored disinformation. The UK also planned to reserve the precious tests for those who were already sick in hospital – a prime example of closing the barn door after the horse had gone. The first UK lockdown was finally announced on 23 March, fully twenty-five days after British health experts had advised it. Having spent the previous weeks shaking hands, Prime Minister Boris Johnson developed symptoms, tested positive on 27 March, and spent a week in hospital, including three days in an ICU. He emerged thinner, chastened, and full of praise for his caregivers. During November, a second UK lockdown appeared more quickly, and a third in January 2021, by which time one hundred thousand of its citizens had died – one of the highest mortality rates in Europe, a position that it held for another year.

Sweden did not implement a lockdown in the first wave, although it too encouraged the basic rules of hygiene and social distancing. Its caseload began to decline in the summer along with other northern countries. However, Swedish mortality in 2020 was high, between five and ten times greater than that of its Scandinavian and Baltic neighbours (table 6.1).

The reasons for Sweden's deplorable statistics are still under analysis. Being able to survive symptomatic COVID-19 depends on access to medical care, oxygen, drugs, fluids, and respirators. If hospitals are overwhelmed, then access to life-saving treatment is compromised. When people in care homes are left untended, they die. Swedish epidemiologist, Helena Nordenstedt, was quoted by the BBC, in July 2020, saying, "If you take care homes out of the equation things actually look much brighter." Was she serious? Did no one care about the elderly? Canada has no reason to feel smug about her shocking, ageist statement.

By the summer, only an estimated 6 per cent of the total Swedish population showed antibodies, herd immunity had not been reached, and

Table 6.1 | Cases and deaths per capita in Scandinavia and the Baltic as of December 2020

Country	Cases per 1M	Deaths per 1M
Sweden	23,875	656
Denmark	13,777	142
Latvia	8,811	102
Estonia	9,124	85
Iceland	15,218	74
Finland	4,463	71
Norway	6,741	62

Based on data posted at CovidNet, 1 December 2020.

the national death toll was embarrassing. Sweden reconsidered its decision in its second wave and implemented a lockdown. Lessons from the pandemic kept coming.

Beginning in March 2020, a group at Oxford University's Blavatnick School of Government developed a *stringency index*, based on nineteen government-response indicators that considered various closures (e.g., schools, businesses, gatherings), supports (e.g., sick pay), and protections (e.g., testing, vaccines), and from them, it generated scores out of one hundred that changed through time according to policy responses to the situation. They kept refining the calculations during the following year, attempting to relate the stringency of lockdowns with cases and deaths. Eventually, they had data sufficient to compare lockdowns within countries and to allow for international comparisons of the effectiveness of response.

Beginning in Asia during the summer of 2020, the metaphorical term circuit breaker was invoked to indicate a short, sharp, partial lockdown, avoiding the negative metaphorical connotations of emprisonment and financial strife. The new term quickly became widespread in the UK media and beyond. Meanwhile, the policies triggered reevaluation of the meaning of the word lockdown and of the concept of freedom from ethicists, legal scholars, and health policy analysts.

The word lockdown, though descriptive and apt, became inflammatory. Some people who lacked social supports equated it with financial collapse. Civil liberty groups, some religious leaders, and some politicians viewed lockdowns, like face masks, as an unwarranted infringement on their freedom – a human rights issue. They did not want a nanny state telling them what they could and could not do. Encouraged by misinformation spewed by President Donald Trump, many Americans (and some Canadians) became convinced that the pandemic was a hoax, nothing worse than a cold, and that if they fell ill, they would survive. They believed that no one should interfere with their choice to accept risk. This cavalier, macho attitude was selfish, in that it posed enormous danger to others, especially the elderly or those with chronic illness, obesity, hypertension, kidney disease, cancer, or immunodeficiency. Displaying their convictions, some flouted recommendations and held enormous, maskless gatherings that became superspreader events.

By October 2020, an international group of epidemiologists, scientists, and physicians (including two Canadians) signed a statement, called the Great Barrington Declaration, named for the Massachusetts town where it originated. They expressed concern over the harmful financial and psychological consequences of lockdowns and school closures and advocated a laissez-faire attitude that would lead to herd immunity.

Also in October 2020, the WHO acknowledged the hardships provoked by lockdowns, but instead of opposing them, it advocated greater supports during isolation periods, contending that herd immunity was best achieved through vaccination rather than infection. It urged countries to work together to "build back better," a phrase originating in the United Nations 2015 plans for disaster risk reduction and adopted by the Biden election campaign.

Debate over lockdowns constantly pitted public-health experts against politicians who feared electoral backlash. Surprisingly, a Canadian signatory of the Great Barrington Declaration, Dr Matt Strauss, became the medical officer of health in Haldimand Norfolk in September 2021. With no public-health credentials, he had repeatedly criticized public-health measures, labelling lockdowns as "unethical," and tweeting (3 August 2021) that he would "sooner give [his] children COVID-19 than a McDonald's Happy Meal." A few months later, however, he was supporting vaccines for kids. Differing control measures, not only between countries, but between and within provinces and regions more often reflected varying political perspectives rather than conflicting scientific views.

Opposition to sanitary recommendations continued in the form of flagrant disobedience. The 6 January 2021 rioters at the US Capitol emphasized their disgust with the rules by deliberately coughing in the faces of elected representatives who were wearing masks. This was not America's finest hour. Canada, too, saw numerous, unmasked protestors blocking access to hospitals, chanting outside the homes of public-health workers, and uttering death threats.

ECONOMICS

Lockdowns may be helpful for controlling a pandemic, but they are not sufficient, and they come with economic costs that aggravate the financial hit owing to the pandemic itself. By November 2020, Argentina with its long, strict lockdown had suffered a case rate equal to that of Brazil (twice that of Canada) and forty thousand deaths. Moreover, its GDP had crashed by a whopping 12 per cent. As this sad example shows, lockdowns simply cannot work if a large proportion of the population lives in poverty, labours in precarious employment, and has no social safety net. People in these circumstances cannot follow the rules.

Historical examples provide reassurance about how economies and societies recover from the hardships wrought by pandemics. One oft-cited but endlessly controversial view is that fourteenth-century plague, with its estimated twenty-five million deaths, overturned the entire economic structure of Europe, helped to end the feudal system, and prompted the rise of a middle class. Some go so far as to suggest (controversially) that plague caused the Renaissance.

Following SARS, economists Steven James and Timothy Sargent of the Canadian Department of Finance showed that its financial impact in both Asia and Canada was sharp but short. They also noted that only a few studies of natural disasters had bothered to compare original predictions with actual outcomes when they eventually became available. In contrast to dire, mid-crisis estimates, data revealed only modest short-term effects on the usual indicators, such as industrial production, business activity, mass transit use, import and export levels, and retail sales. Using these indicators, they showed that the economic recovery following the 1918 influenza pandemic had been more rapid and robust than predicted at the time.

Early in COVID-19 and likewise referring to 1918 influenza, a team of American economists compared economic recovery in countries that had taken strict public-health measures against those that had not: the former managed to build back much better than the latter, confirming the earlier analysis of James and Sargent.

But something new had appeared following World War I and the 1918 influenza pandemic. It was the notion of a welfare state and the idea, attributed to economist John Maynard Keynes, that governments owe their citizens support in a crisis and should tolerate deficits, thereby helping to revive the economy when the crisis is over. The benefits sustain people's lives during a catastrophe and beyond to stimulate the future economy. They might include direct cash payments to individuals or businesses, tax relief, forgivable loans, and investment in repairing damages – whatever they may be. Relief benefits were used following World War II, and again, after a period of unpopularity, they made a big return during the global financial crisis of 2007 to 2009. They are sometimes called stimulus packages.

COVID-19 generated an immediate and deep recession. Nonessential businesses were closed. Many manufacturing concerns came to a standstill; the few exceptions included food processing and the production of (and frantic search for) PPE. Firms that had not been forced to close eventually did so voluntarily, because they could not obtain parts or sell products. The pharmaceutical industry confronted increased demand but had difficulty obtaining raw materials, shipping finished products, and keeping plants open and staffed. Airline travel went into steep decline. Some nations temporarily or permanently banned flights from certain regions known to have high rates of infection. Others, like Australia, closed their borders altogether and grounded all flights except for emergencies. Airline companies suffered deep financial losses as passengers decided not to fly for fear of being trapped inside a sealed metal tube with potential carriers seated nearby.

By March 2020, world stock markets were down 25 per cent from January highs; the drop was 33 per cent in G7 countries – far greater than had been observed in the first month of the financial crisis twelve years earlier. To counteract the new recession, many governments, including the United States and Canada, stepped up with benefit packages intended to rescue their citizens and economies from hardship (chapter 2). These packages were intended to allow citizens to weather the lockdowns, support health care, and fund the scientific research needed to fight the scourge. No one could tell how long it would last and what more investments might be needed. Nevertheless, the initial relief packages for COVID-19 in the large economies were greater than they had been for the 2008 financial crisis, with exception of China (figure 6.2).

The US *Coronavirus Aid, Relief, and Economic Security Act* (CARES), signed into law on 27 March 2020, provided US$2.2 trillion for various

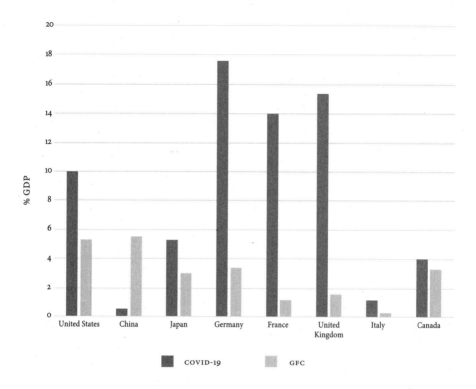

Figure 6.2 | Comparison of relief packages: the 2007–09 financial crisis and early COVID-19 (Source: https://theconversation.com/coronavirus-comparing-todays-crisis-to-2008-reveals-some-interesting-things-about-china-132147, 31 March 2020. By permission of the author, S.M. Chaudhry.)

disbursements, including $1,200 for adults and $500 per child up to $3,400 per household with additional amounts in the event of infection. It was the largest stimulus package in American history, equal to 10 per cent of the GDP, although it was a per capita amount less than initial packages in the UK, France, and Germany.

China too lavished funds to support its citizens affected by the lock-downs. For example, according to a 23 March report in *Vox EU*, China had allocated about US$17 billion to fight the epidemic itself, while the central bank released US$221 billion in local government bonds, at least half of which went into rent, wage, and social security supports with tax relief and stimulus funding to sustain small businesses and pandemic manufacturing, such as PPE production. But these amounts were proportionately less than in the financial crisis; with China's burgeoning economy and localized needs, the COVID-19 disbursements formed a smaller proportion of GDP.

The International Monetary Fund (IMF) and the World Bank also poured money into the coffers of emerging economies in Asia, Africa, and Latin America where margins of support were narrow or nonexistent. All the packages were revisited and revised as the pandemic wore on. These measures were an essential form of pandemic management, but they were controversial. Those opposed to the payments, including politicians given to touting fiscal responsibility, together with citizens who tended to blame the poor for their own misery, had great difficulty comprehending how relief spending could be a form of pandemic control. For them health care was nothing more than modern drugs and vaccines. Even some supporters of benefit payments had reservations, voicing renewed concerns about coloniality in the structures used to deliver relief.

7

Treatments

Unlike bacteria, viruses have been notoriously tricky to treat with specific drugs. Since the mid-twentieth century, we have had antibiotics to manage bacterial infections, with newer agents appearing all the time. But antibiotics have no effect on viruses. That is why, in flu season, doctors resist prescribing them and, instead, encourage rest, fluids, and over-the-counter drugs to reduce pain and fever, improving symptoms while the body heals itself. Sometimes a severe virus infection will invite bacteria to invade damaged tissue; then antibiotics should be used. This situation was probably a major cause of death during the 1918 influenza pandemic before effective antibiotics had been discovered. Doctors also worry that using antibiotics when they are not indicated will simply help to create more resistant strains of bacteria: superbugs.

When COVID-19 first appeared, it produced a viral pneumonia – with symptoms of what is called adult respiratory distress syndrome (ARDS). Recognized since the 1960s, ARDS consisted of a triad of findings: shortness of breath, X-ray evidence of infiltrates in both lungs, and low oxygen levels. The best treatment was not curative but supportive: rest, fluids, fever reduction in mild cases. If lung involvement is extensive, patients might need extra oxygen, inhaled through a mask or nasal prongs. Should exhaustion set in, more help with breathing could come from a ventilator, a machine that uses positive pressure through a tube in the trachea to deliver air and oxygen directly into the chest. The tube is uncomfortable; patients cannot talk, nor can they move much without assistance. Consequently, intubated patients are sedated to the point of an induced coma, where they must remain until their lungs clear, and they have strength enough to breathe on their own. If the tube must be used for a long time, such as ten days to two weeks, it begins to erode the tissues of the throat, and a tracheostomy (hole in the neck) can be performed to maintain and protect the airway. This supportive care is how most patients who are seriously ill with COVID-19 are managed – provided they are lucky enough to find a place in hospital.

The most severe infections trigger a cytokine storm, which, as mentioned in chapter 1, is a massive immune reaction affecting multiple organs in the body and producing widespread inflammation and clots in tiny blood vessels. All too frequently it hails the point of no return as the whole body shuts down; however, it can be managed with drugs that dampen the immune response. The concept of cytokine storm has been recognized in the immunology literature since at least the mid-1990s, but the scant publishing on the topic using the term as a keyword multiplied fourteen-fold from about 130 peer-reviewed articles in 2019 to about 1,800 in 2020, and more than two thousand in 2021.

As familiarity with the symptoms of COVID-19 increased, certain medications were soon examined for possible effectiveness. Many of the recommended prescription drugs were already licensed and found in formularies for other purposes. Sales, hoarding, and profiteering exploded. People who needed these medications for other purposes were often unable to obtain them. Some unusual products had distinguished backers who based their celebrity endorsements on reasoning about the mechanism of action. Other contenders came from the over-the-counter or folk-medicine branches of the industry. Many were falsely proclaimed to be preventions or cures. With almost no scientific evidence, the prime minister of Madagascar began promoting and distributing a drink, called COVID Organics, derived from the artemisia plant (like the antimalarial, for which Tu Youyou shared the Nobel Prize in 2015); however, a trial from Nigeria released in July 2020 proved it was ineffective.

Hydroxycholoroquine is a good example of a highly promoted remedy. This old, reliable, off-patent drug is used to treat and prevent malaria and to manage autoimmune conditions, such as rheumatoid arthritis and lupus erythematosus. Its mechanism of action is related to blocking acidification. Since acidification is part of how a virus invades cells, some scientists reasoned that it might work against SARS-COV-2. In mid-March 2020, the maverick, infectious-disease specialist, Didier Raoult, based in Marseille, announced that he had been inspired by an earlier Chinese trial and had himself obtained good results in twenty-four patients with COVID-19. He advised using it with the antibiotic azithromycin. France promised more investigations.

Later in March and on no other evidence, President Trump touted the benefits of hydroxychloroquine, claiming he was taking it himself. The National Institutes of Health (NIH) expert, Dr Anthony Fauci, disagreed with the president. Within days and continuing for months, dense shortages of hydroxychloroquine were reported in the United States and globally, especially in Asia. By April, various American physicians argued that the severity of the COVID-19 crisis meant that anything, which might be effective, should be tried. In this manner, cardiologist Peter A. McCullough became a proponent of hydroxychloroquine in March 2020 and led an April 2020 study on its effectiveness in Texas. With others, he proposed an early treatment regimen. McCullough became a regular critic of public health in right-wing media, such as Fox News. He addressed a US Senate committee about suppression of information, but was eventually accused of spreading misinformation, touting natural immunity over vaccines, and distorting evidence. By July, his former employer, Baylor Scott & White Health, sued him for persisting in using his Baylor University affiliation for more than six months after it had come to an end. Although the lawsuit was over his self-styled affiliation, not his statements on COVID-19, for skeptics in the Trump-drenched atmosphere, the lawsuit made him a victim, a maligned whistleblower who had dared to speak truth to power and was being punished for it.

In April, a rheumatologist in my city told me that his patients with arthritis and lupus were struggling to find their regular hydroxychloroquine medicine and had to reduce the dose to stretch supply. To his disgust, he had even been approached by physician colleagues asking him for a personal prescription for the drug to use in case of need. In May, the UK government bought millions of doses of hydroxychloroquine to test and stockpile, pending results from the trials. Yet at the same time, France revoked the decree that had authorized its off-label use for COVID-19 because new trials revealed its dangers, especially when combined with the antibiotic azithromycin, as Raoult recommended. The popularity of the unproven remedy persisted even after warnings came that it could (and did) lead to irregular heartbeat or cardiac arrest.

Several robust trials eventually confirmed that hydroxychloroquine did more harm than good in preventing or treating COVID-19. One of the most convincing trials was a natural experiment conducted on a large group of people taking the drug for other diseases: they were found to be just as vulnerable to COVID-19 as age-matched controls not taking the drug.

Similar reasoning about plausible mechanisms of action drew attention to a host of other remedies, some from the realms of herbal medicine and dietary supplements. Several doctors, including the controversial, South-African-born American, Paul E. Marik, eschewed clinical trials, claiming that the crisis demanded immediate action: he reasoned that because people are dying, the rules should be skipped. Not far behind, lurked entities seeking to profit from the demand for any potentially helpful remedy. In the United States and elsewhere, prices for these readily available commodities soared, and they too became scarce. Investigators around the world launched a flood of in-vitro studies (conducted in a laboratory, not in a human) to assess hypotheses about these remedies: zinc, garlic, nicotine, vitamins, and melatonin.

As early as March 2020, some scientists were recommending that *Vitamin D* be studied for preventing and treating COVID-19. Already known to be helpful in preventing respiratory illness and enhancing immunity through its effect on calcium, it is made in our bodies upon exposure to sunlight. In the winter seasons, when sunlight levels are low and people stay inside, its level falls. Much chatter surrounded the possibility of its immune-boosting capacity, and in September, Anthony Fauci admitted that he was taking it himself. In October 2020, a study out of Spain claimed that 80 per cent of hospitalized patients with COVID-19 had much lower levels of Vitamin D than a control group. The authors made no claim about its impact on the duration of disease or its severity. By January, the UK was still on the fence about its value, but it began providing low-dose Vitamin D supplements to vulnerable populations in care homes. Won't hurt; might help. Most observers contended that doses were too small. In June 2021, researchers at McGill University published a large study suggesting that Vitamin D levels made no difference to COVID infection and outcome.

Because hundreds of very sick people were being hospitalized, it soon became possible to compare the effect of various standard treatments. The reliable old steroid, *dexamethasone*, has been around since 1957, used often in cancer care and in dealing with severe inflammatory disease and brain swelling. It is a synthetic and more potent derivative of natural hormones from the adrenal cortex, the discovery of which brought the 1950 Nobel Prize to P.S. Hench and E.C. Kendall. Use of dexamethasone for inflammation of the lung might seem to be an obvious choice; however, one of its side effects is to reduce immunity – possibly increasing susceptibility to infection. No one knew if it could help in COVID-19 or not. In mid-July 2020, a UK consortium published a large, randomized trial of more than two thousand patients in the *New England Journal of Medicine* showing that dexamethasone significantly reduced mortality in COVID-19, but only for patients who were already seriously ill on ventilators. This discovery altered the standard of care: now advocating dexamethasone for those in dire circumstances and protecting others from an unnecessary drug and its side effects.

Later in the pandemic on 22 January 2021, a group at the Université de Montréal announced good results obtained with the old drug *colchicine*. Derived from the autumn crocus, the plant has been used since antiquity to treat the inflammation of gout. It was isolated (purified) in the early nineteenth century and is now known to inhibit cell division (mitosis). In COVID-19 patients, it reduced hospitalizations by a quarter, and intubation and death by almost half. Funded by the Gates Foundation and the Quebec government (among others) and conducted in several countries, the randomized, double-blind, placebo-controlled study had several revolutionary aspects: it entailed an inexpensive oral medication taken by patients in their own homes as soon as they received a diagnosis. The reasoning that motivated its use was related to the lethality of the cytokine storm in COVID-19 and the observations that the original SARS virus activates an inflammatory mediator, called an inflammasome. Because colchicine is known to reduce activity of inflammasomes, the researchers hypothesized that it might reduce complications of COVID-19 by preventing or reducing the inflammatory syndrome. It appeared that they were right, and the standard of care changed again, this time for patients who were not very ill. Shortages were predicted, but a year later, conflicting reports over the value or failure of colchicine were still appearing almost weekly.

Similarly, news came out of Toronto in late January 2021 that treating moderately ill patients with the old, reliable, blood thinner, *heparin*, had a beneficial effect. Originally discovered at Johns Hopkins University, studied in Sweden, and purified in Toronto between 1929 and 1937, under the leadership of Charles Best (of insulin fame), heparin is one of the most widely used treatments for clots in human limbs or lungs and for preventing clots in people on hemodialysis or other machines. It is also applied to the extremely dangerous condition of DIC (disseminated intravascular coagulation), which occurs in those who are severely ill with sepsis or shock, sometimes after childbirth or snakebite. Under those circumstances, people bleed profusely because their illness has triggered microscopic clotting throughout their bodies and their clotting proteins are used up. Paradoxically, a blood thinner, like heparin, can reverse the process. It sometimes takes courage to give it to a bleeding patient. In any cytokine storm, a DIC-like process can take place, and since at least June of 2020, clinicians had reported evidence of DIC in serious cases of COVID-19. Giving heparin in advance might prevent it.

Chief among the licenced drugs being repurposed against COVID-19 was *ivermectin*, originally isolated in 1970, from streptomyces bacteria found by Irish-American William C. Campell and Satoshi Omura of Japan. Since 1981 it was known (and approved) as an effective agent against several parasitic worms in humans and animals under trade names Mectizan or Stromectol. Its maker, Merck, began donating it in 1987 to eradicate river blindness and other parasitic diseases in Africa, Yemen, and Latin America. The program, which continues, has seen many successes and served as a model for other donation programs. Each year, three hundred million people are treated, and more than four billion doses have been donated. The science behind ivermectin – a drug derived from bacteria – resembles the work of Selman Waksman, already awarded a Nobel Prize in 1952. Consequently many regard the share of the 2015 Nobel Prize awarded to Campbell and Omura as recognition, not only for their discovery of ivermectin and its prowess against neglected diseases, but also for the impressive philanthropy associated with its use.

With the advent of COVID-19, results from Monash University in Australia suggested that ivermectin might help because in high concentration it inhibited SARS-COV-2 virus growth in monkey kidney cells in vitro. A paper released before peer review in November 2020, by Ahmed Elgazzar and colleagues in Egypt, claimed that it provided an almost complete reduction of morbidity and mortality in four hundred COVID-19 patients compared to controls. Soon people around the world pleaded with caregivers to provide the unproven remedy. But two people in South Africa ended up in ICU having overdosed themselves. Stocks were depleted. People desperate to acquire ivermectin resorted to ingesting or applying the off-the-shelf veterinary preparations, including ointments. By early 2021, its benefit was still considered unproven, and clinicians coping with a severe caseload in Lagos, Nigeria, announced that they were launching a clinical trial, while others in Israel reported reduction in viral shedding in mild cases.

Some suspected that Merck, the manufacturer of the only human preparation of ivermectin licensed in Canada, was uninterested in restoring ivermectin supplies or supporting an expensive clinical trial because it would interfere with future sales of its new and potentially more lucrative, oral antiviral, molnupiravir, which promised to earn billions by 2022 (below). Even so, by mid-January 2021 the human and veterinary versions of ivermectin fell into short supply in Canada for the first time with no anticipated return of availability until mid-2022.

Ivermectin soon garnered an eclectic group of supporters, including respected expert in evidence-based medicine, Dr Tess Lawrie. She founded the British Ivermectin Recommendation Development (BIRD) Group in early January 2021 after hearing American critical care physician, Pierre Kory, tout its benefits. On 9 February 2021, researchers at McMaster University proposed to participate in an international trial on ivermectin (with metformin and fluvoxamine) with funding from the Gates Foundation; however, the Canadian bureaucratic hurdles were so great that no participants were to be enrolled in Canada. Many other trials were launched, and the American courts got involved. In mid-2021, an Ohio judge ordered doctors to give ivermectin to a patient whose family wanted it – a decision reversed a few days later. In Virginia, Dr Paul Marik sued his employer for the right to give ivermectin to his patients, but he eventually resigned his position.

Also in February 2021, other Canadians got involved with ivermectin, including family doctor Ira Bernstein in Toronto who had seen good results anecdotally, retired pharmacologist Kanji Nakatsu from Queen's University who understood the presumed mechanism of action, and New Brunswick chartered accountant David Ross. On 25 March after appealing numerous times to government by letter, Nakatsu launched a House of Commons petition for greater access to human ivermectin, garnering more than the requisite five thousand signatures within thirty days. It was presented in the House in June by Conservative member of Parliament Dean Allison; however, it remained unanswered at dissolution of Parliament with the election call in mid-August 2021. Nakatsu then revived it. For his part, accountant David Ross oversaw and became a director of the non-profit Canadian Covid Care Alliance (CCCA), incorporated 23 April 2021 and devoted to questioning established science. Boasting five hundred members, the CCCA and its various committees began holding regular meetings virtually. Members include Paul Elias Alexander who holds a doctorate in health research and had worked in the Trump administration where he attempted to silence CDC scientists by delaying their weekly publications and editing their pandemic statements to be more upbeat and compatible with the president's political utterances.

Some countries went ahead and licensed ivermectin for off-label use: the Czech Republic, Mexico, Peru, India, although the latter two later rescinded the decision. Advocates correlate decline in national caseloads with its use. Meanwhile, Lawrie's BIRD group and Kory continue to argue for more research and greater access to the drug. With suspicion that Big Pharma suppression by Merck might exist, other conspiracy theories attached themselves to the agenda, and ivermectin promotion everywhere became barnacled with promotion of other unproven remedies, COVID-19 deniers, and antivaccination protesters. Many of its supporters, like CCCA members, were serious scientific researchers who understood the *plausibility* of the hypothesized mechanisms of action and were wary of platitudinous claims and edicts flowing from governments or Big Pharma. The absence of effective treatments was another driver. Nakatsu signed his email messages, "While we dither, people die." The CCCA website links a page, called Vaxxtracker, which boasts that it is a "safe-space" for vaccine recipients to "report negative side-effects." CCCA soon became a platform for advocating and locating a multitude of other unproven remedies, posting criticism of accepted treatments and vaccines, and finding work-arounds for people wishing to be exempt from vaccines or lockdowns. These affiliations did not help ivermectin.

In July 2021, a British graduate student and misinformation blogger, Jack Lawrence, reported serious flaws in the original Elgazzar paper, which led to its retraction. Then establishment opinion, which had been tolerant, turned hostile. Health agencies, physicians, and even the WHO piled on, warning about the dangers of ivermectin, especially in veterinary preparations. On 21 August, the US FDA resorted to humour and tweeted: "You are not a horse. You are not a cow. Seriously y'all, stop it!"

Ivermectin supporters found deep irony in the fact that a relatively harmless drug, which had won the Nobel Prize, had been given safely in billions of doses each year (for parasites), and cost only pennies, would be so demonized, while, for example, novel products sailed through approvals despite small trials, minimal effectiveness, and astronomical price tags. That irony, compounded by the fact that human preparations of ivermectin were utterly unavailable, enhanced their determination to see decent studies through to completion.

At the time of writing (2022), CCCA and BIRD are still active and the registry of clinical trials around the world lists eighty-three studies on ivermectin for COVID-19: thirty-one are still recruiting, eighteen are not yet recruiting, thirty are completed, three have been terminated, and one has been withdrawn. It is scarcely surprising that seventeen of these studies are based in Egypt, eager to regain its credibility in evidence-based research.

Another compound that received far less attention but had stronger associations with actual antiviral activity was *quercetin*, a plant-derived substance found in many fruits, vegetables, and seeds, including red onions, capers, and kale. It had been an over-the-counter remedy sold as an antioxidant in health food stores for several years. Following SARS, its in-vitro properties against a host of viruses were explored by many groups, especially in China and Canada, including that of the distinguished scientist, Michel Chrétien (brother of the former Canadian prime minister) and his Congolese-Canadian colleague, Majambu Mbikay, both of Montreal. In 2014, working with the infectious disease laboratory in Winnipeg, they published a trial demonstrating its effectiveness against Ebola virus in mice in 2016. In late February 2020, *Maclean's* magazine ran a feature article on Chrétien's research and his longstanding experience collaborating in China. His laboratory received a C\$1 million donation from the Lazaridis Family Foundation to test quercetin against novel coronavirus infection, although at least five times that amount would be needed to cover costs of running convincing trials. Originally planned for China, by the fall, the trial was moved to Canada where the outbreak had grown more extensive. At the end of 2020, more than a hundred institutions were speculating on, or conducting research on the use of quercetin and similar flavonoids in preventing and treating COVID-19. Available off-the-shelf in Canadian grocery and health-food stores in early March, the oral form soon disappeared from stores and mysteriously vanished from the Health Canada Drug Products Database, although it can still be found for sale online.

Another set of old drugs, already licensed for mental health, drew attention for their potential to modulate the cytokine response. In mid-2020, studies done in people and in laboratories from France, Brazil, and the United States, suggested that *fluvoxamine* – an antidepressant – would reduce the risk of severe disease and hospitalization if taken as soon as symptoms appeared. Officials were on the fence about its value, but the late massive surge in the pandemic caused them to reconsider. Ontario was one of the first to grant tentative approval in January 2022.

ANTIVIRAL DRUGS

Over the last half a century, some *antiviral drugs* have emerged with mixed results. They aim to interfere at various points in viral structure, life cycle, reproduction, or means of invading cells. One of the first was acyclovir, which I remember using in attempts to stave off severe forms of shingles in the late 1970s. At least two dozen others – the antiretrovirals (ARVs) – were developed to block HIV or the human enzymes that the HIV uses to replicate itself; great strides were made with these drugs in managing AIDS. Readers may also be familiar with oseltamivir (Tamiflu), an antiviral developed by the Gilead company in the late 1990s. Derived originally from the star anise plant, it is effective against some forms of the influenza virus, but it has a very short shelf life. From 2013, the Gilead company also patented several drugs (sofosbuvir; ledipasvir, etc.), which constitute an almost miraculous antiviral therapy for hepatitis C, especially when used in combination. But the cost of a course of treatment, at US\$100,000, was exorbitant. Gilead worked with generic companies around the world to reduce costs even though its various American patents would not expire until the 2030s.

Among the bouquet of drugs that Gilead had developed to combat hepatitis C, was the drug *remdesivir*, first created in 2009 and known as GS-5734; however, it did not work for hepatitis C, and was displaced by more effective new drugs. It was then repurposed and studied as a possible treatment for other diseases, including Ebola. By 2017 it was found to have activity in vitro against other viruses, including coronaviruses, among them those causing SARS and MERS. Gilead was granted a US patent for its use against coronaviruses in April 2019, only eleven months before the pandemic. That same month, two researchers from the US Army Medical Research unit concluded their review of coronavirus therapies, including GS-5734, with these chillingly prophetic words: "Based on lessons learned from SARS and MERS outbreaks, lack of drugs capable of pan-coronavirus antiviral activity increases the vulnerability of public-health systems to a highly pathogenic coronavirus pandemic." As I keep stating, we were warned – not only of the coming pandemic but also of the staggering lack of effective treatment.

Obviously, as the COVID-19 pandemic began, caregivers threw whatever antivirals they had available against the new virus, chief among them the newly patented, possibly specific remdesivir, because it had proved effective against some coronaviruses. In the heat of the crisis, the usual caution in planning randomized controlled trials was barely possible – while the drug's limited availability and astronomical price it put beyond reach for most hospitals. By spring 2020, publications emerged of promising trial results. One of the first, published in the *Lancet*, concerned 236 patients in Hubei province. Another – a high quality, randomized, placebo-controlled, blinded investigation, published in the *New England Journal of Medicine* – had recruited over a thousand patients at sixty sites on three continents between February and April 2020. These studies confirmed that remdesivir could help, but only to reduce the duration of illness in the severely ill. Furthermore, serious side effects were noted in many patients receiving either the drug or the placebo; at the time of writing, the actual rate of dangerous side effects is unclear. Nevertheless, already the subject of great interest, nations grew avid to obtain remdesivir and shortages were widely reported, although the drug had never been available, or even desirable, until COVID-19 made its prospects soar. Gilead set a greatly reduced price for developing countries, which some observers recognized as reputational repair needed because of the prohibitive prices it charged for its successful hepatitis C product.

Other antivirals, including favipiravir, used in Japan for influenza, and the combination of lopinavir and ritinovir, used for HIV, have been brought into clinical trials against COVID-19 so far with unconvincing results.

Most antivirals require injections, but in October 2021, Merck's long-awaited oral antiviral, molnupiravir, had completed trials. The pharma giant applied for approvals and announced that it would allow generic companies in poor countries to make and sell it at low cost. Merck was thereby planning to enhance access equity, just as it had done with its original distribution of ivermectin. In some trials, molnupiravir performed little better than placebo, but by late 2021 it was approved for use in several countries including India and the US. Pfizer also developed a new antiviral drug, a combination of nirmatrelvir and ritonavir (Paxlovid); like its Merck competitor, it too could be taken as a pill and received approval for emergency use from the US FDA on the same day. Health Canada granted approval on 17 January 2022, while a decision on the Merck product was still pending. Short supply meant that only qualifying patients, such as those with immunodeficiency, could receive the new oral antivirals.

By December 2020, the WHO had published the results of a trial on four drugs – remdesivir, hydroxychloroquine, lopinavir, and interferon – involving more than eleven thousand patients from 405 hospitals in thirty countries. The endpoint was mortality. The news was disappointing: none of the drugs made any difference to survival. Nevertheless, the meticulous work and collaborative engagement to conduct such fine studies rapidly in the overwhelming chaos of the pandemic was exceptional and hopefully will leave a legacy for future research of any kind.

ANTIBODY TREATMENTS

Early on, historians of medicine began to wonder if and when studies would be launched on treating the sick with the natural antibodies generated by people who had survived COVID-19. Bert Hansen published a review of what is called serum therapy in *Distillations* of April 2020. This method of passive immunization dates back more than a century to the 1890s when animal serum, loaded with antibodies against tetanus or diphtheria, was used to treat humans. The first Nobel Prize for medicine in 1901 went to Emil von Behring who had used an antitoxin serum against diphtheria, a technique that relied, not on the recipient's own ability to generate antibodies, but on antibodies borrowed from an immunized animal.

But the method of using convalescent plasma from humans or animals who had survived illness fell into decline with the advent of antibiotics in the 1940s; it seemed more logical and attractive to kill the invading organism directly and to avoid possible contamination with other infectious agents or allergic reactions to foreign proteins. In the intervening decades, however, technical advances meant that passive immunization might offer an effective and relatively safe pathway. The advances included new methods for testing and purifying blood products, for collecting specific blood components, and for generating biological pharmaceutical agents, called monoclonal antibodies, known by their suffix "-mab."

Sure enough, whispers of anecdotal evidence of transfusion benefits and possibilities began appearing in small studies out of China in March 2020: five patients so treated survived with a short illness; ten patients tolerated the method well; six patients demonstrated high levels of antibodies against the virus; six patients with COVID-19 tolerated the convalescent plasma infusion, recovered quickly, and cleared the virus. Convalescent plasma would contain not only antibodies against SARS-COV-2, but also many other antibodies and proteins. And, like any transfusion, the plasma would have to be matched for blood type.

Scientists were looking for antibodies that were pure and specifically directed against the virus. Could readily available biologic agents work like the transfused antibodies? Some worried that in a cytokine storm, biologics that act against the immune response might make things worse by reducing natural defences against the infection itself.

The biotech pharmaceutical industry was already on the case, looking for streamlined equivalents of the kind of antibodies made by COVID-19 survivors. As early as 4 February 2020, Regeneron, an American biotech company founded in 1988, announced it would begin elaborating monoclonal antibodies to fight the pandemic. When the Trump administration announced its Operation Warp Speed in mid-May, Regeneron became one of the beneficiaries with US$450 million to make and supply an antibody cocktail aimed at both preventing and treating the disease. The drug, called REGN-COV2, contains two different monoclonal antibodies (casirivimab and imdevimab) designed to block viral infection and reproduction. It was still in clinical trials in October 2020, when the president contracted COVID-19 and was treated with it via a compassionate-use request. Perhaps the recently recovered POTUS was still cruising on a dexamethasone high when he announced on 7 October that REGN-COV2 was a "cure" that made him better "immediately" and should be provided for free to everyone, because the pandemic was "China's fault." Although trials were not complete, FDA granted approval for emergency use in November. Costly and complicated to make, the drug is (and was) not widely available, although the celebrity endorsement made it widely in demand. The US government was distributing it for free in 2021, until it proved ineffective against the *omicron* variant. In late January 2021, Germany spent half a million Euros for two hundred thousand doses.

Again, initial evidence was anecdotal, and it came from the natural experiment of people taking such drugs for other conditions. In the summer of 2020, rheumatologists in Spain and dermatologists in Italy observed that their patients who were using biologics (e.g., dupilumab, adalimumab, and inhibitors of either interleukins or tumour necrosis factor) for autoimmune diseases, including arthritis and psoriasis, seemed protected or had a better outcome than their neighbours when confronted with COVID-19.

Another MAB drug that was investigated in published trials at least twenty times across 2020 was tocilizumab (Actemra) widely used since 2010 for juvenile arthritis and other autoimmune diseases. Because it was already indicated for treating cytokine storm, a study published in the *New England Journal* made the following tepid conclusion: "Patients who received tocilizumab had fewer serious infections than patients who received placebo ... [but] tocilizumab was not effective for preventing intubation or death in moderately ill hospitalized patients with Covid-19. Some benefit or harm cannot be ruled out, however, because the confidence intervals for efficacy comparisons were wide." These cautious results went viral. With the rise of the aggressive *delta* variant of SARS-COV-2 in the summer of 2021, dense, global shortages of the drug provoked serious harm for children and adults who had been relying on it to control their arthritis. Shortages owing to increased demand began in Canada in late 2021.

At least seventy different antibodies are under investigation, a number of them made in Canada in collaboration with other pharmaceutical firms and financial aid from the taxpayer (never considered enough by would-be developers). One such candidate is bamlanivimab, developed by Vancouver-based AbCellera Biologics in partnership with Eli Lilly and approved for emergency use in high-risk patients by the US FDA, together with REGN-COV2, in early November 2020 and by Health Canada one month later. The FDA authorization was revoked in April 2021 because the drug was deemed ineffective against new variants of the virus.

The COVID-19 crisis provided an opportunity for the pharmaceutical industry to address the long overdue need to reconsider some of its competitive practices. Duplication of research endeavours was a waste of time. Furthermore, the unprecedented demand for vaccines strained the capabilities of factories, inviting the players to consider more collaboration, which slowly grew during the second wave and with the advent of vaccines.

Well before the pandemic, Canadian scientists participated in several open-science initiatives, that is, without patents. For example, the international public-private Structural Genomics Consortium (SGC), founded in 2003, aims to accelerate the development of new drugs using open science. Its spin-off, for-profit companies do not file patents and, since 2017, are wholly owned by a Toronto-based charity, Agora Open Science Trust. Similarly, in May 2020, prompted by the pandemic, the open-science Viral Interruptions Medicines Initiative (O-VIMI) was launched by a Montreal-based consortium of academics, health-care providers, and government, creating an international network of interlocked nonprofits to identify drug targets, develop drugs against viral families with pandemic potential, and test them in early clinical trials. These new drugs are to be created in what O-VIMI calls "a public good." They will eventually sit ready to be rapidly deployed to treat COVID-19 or a future pandemic. The website features a Chinese proverb that applies well to all forms of pandemic planning: "the best time to plant an oak tree is twenty years ago; the second-best time is now."

Figure 7.1 | Editorial cartoon by Dave Whamond, *Washington Post*, 2 July 2020 (Courtesy of Dave Whamond and Cagle Cartoons.)

After this ramble through the tangled weeds of therapeutics, it should be abundantly clear that – so far – there are no cures for COVID-19. The best measures for confronting this disease are supportive treatment and the ancient, tried-and-true gestures of prevention (figure 7.1).

8

Vaccines

The unprecedented speed with which effective vaccines were created for COVID-19 is one of the most amazing features of this pandemic history. For more than two centuries, vaccines have been an established part of disease prevention; however, they have often proved elusive. Although they are much desired, effective vaccines for major diseases, such as HIV-AIDS, do not yet exist; some are only partially effective. After more than fifty years of struggle, a vaccine for malaria was approved in late 2021. At the beginning of the COVID-19 pandemic, all experts insisted that the average time to make a vaccine – if one could be found – was more than a decade.

BACKGROUND

As soon as the genetic code of the SARS-COV-2 virus was known, laboratories around the world began working furiously on vaccines, and media sources began pointing to them as the eventual salvation. For prevention, vaccines are *active* immunization, which stimulates the host's own immune system, as opposed to the *passive* form, which comes from an actual infection or from serum therapy with some other creature's antibodies (chapter 7). In antiquity, survivors of the fifth-century BC plague of Athens were known to be safe from re-infection; they had been protected by what we would call a form of passive immunization (often called natural immunity).

By at least early modern times, some parents took to allowing their children to catch what they hoped would be a mild form of an ambient disease to make them resistant to worse varieties that might come along later. Evidence suggests that it was practised even earlier in Africa.

In the eighteenth century, smallpox was widespread across Europe, causing much suffering, disfiguration, blindness, and many deaths. In 1717, the British ambassador's wife, Lady Mary Wortley Montagu, observed citizens of Constantinople inoculating their children with matter taken from the sores of milder cases of smallpox, also called variola. She became a crusader for the process that came to be called variolization and had her own son inoculated.

Later in the century, a milkmaid taught the surprised doctor Edward Jenner that people who had once had cowpox would never catch smallpox. This folk wisdom, familiar to milkmaids, was of unknown origin and antiquity. Cowpox is also called vaccinia, from Latin *vacca* for cow. In 1796, Jenner experimented on eight-year-old James Phipps, inoculating him with cowpox taken from a blister on the arm of dairy maid Sarah Nelmes (hence vaccin[ia]ation). Then, two months later, Jenner challenged him with smallpox. Young Phipps did not develop smallpox. Our horror at the appalling ethics of this experiment is softened by the realization that Jenner's smallpox challenge was variolization, as recommended by Lady Mary and used on her offspring and those of several of Europe's crowned heads for decades.

Some people were understandably disgusted by the unnatural thought of injecting fluid from a sick animal or human into the bodies of their children or themselves. An antivaccine movement sprang up almost immediately, and cartoonists, like James Gilray, lampooned the procedure, depicting vaccine changing people into monstrous cattle.

Nevertheless, news of Jenner's discovery was rapidly communicated, and supplies of vaccinia were widely distributed. In 1807, he sent a gift of vaccine to the Indigenous people of Fort George, Canada. In the nineteenth century, special farms provided a steady supply of vaccinia by raising deliberately infected cows and horses too. For the next 150 years, smallpox vaccine protected populations in many countries, usually on an ad hoc basis whenever outbreaks occurred. In one of its first campaigns, the WHO committed to the eradication of smallpox: combined with careful identification, isolation of cases, and quarantine of contacts, vaccine allowed it to announce success in 1979. Smallpox was the first and, so far, the only human disease to be eradicated. Variola virus still resides in American and Russian laboratories, and its eventual destruction is a matter of debate.

Using similar principles, scientists found ways to prevent other diseases through active immunization, all of which are now called vaccination, as a general term, although they have nothing to do with vaccinia virus or cows. Louis Pasteur developed a vaccine against rabies with a weakened strain of the virus, first used successfully in humans by 1884. Von Behring's 1901 Nobel Prize was in part for work concerning diphtheria immunity.

Not every infectious disease could be countered with a vaccine, although attempts were made. In the experimental stage, vaccines sometimes provoked the disease that they were trying to prevent, as was the case for syphilis. Vaccinia sometimes resulted in side effects, such as severe eczema or even meningitis, although it protected against smallpox. As the WHO eradication campaign neared its end and small-pox waned, routine vaccine for smallpox was no longer recommended, the risk of side effects having become greater than the risk of disease. Vaccines against malaria, cholera, and tuberculosis are only partially effective, but others – against viruses causing yellow fever and measles – afford remarkable immunity for life.

In 1955, polio vaccine was licensed. It contributed to better control of the regular outbreaks of this paralyzing disease, which, thanks to a similar WHO campaign, is expected to become the next human disease eradicated from the globe. Diphtheria, whooping cough, and tetanus vaccines were also recommended in a combined shot. Over the next seventy years, new vaccines and improvements in existing vaccines emerged and the immunization recommendations kept changing to recognize these achievements.

Influenza vaccines were slow to appear because the virus mutates easily, which is why we get flu shots yearly. The first influenza pandemic in which vaccines played a promising role was the 1957 so-called Asian flu. By that time, the scientists were able to manipulate the vaccines, making them better able to target the invading virus.

As a kid in the 1950s, I suffered both measles and chickenpox and I received mandated vaccines against smallpox and tuberculosis. People my age also remember being sent out to play with neighbourhood children who had mild chickenpox. The idea was to get it over with, just as people had done for smallpox in earlier times. But now vaccines for measles, mumps, rubella, whooping cough, haemophilus influenzae, and HPV (cervical cancer) are available as part of the immunization schedules for children. Vaccines for hepatitis, pneumococcus, and chickenpox (or shingles) are also available. Nevertheless, over the course of a half century, pharmaceutical companies began to abandon vaccine production, because they were expensive to research and manufacture, and the market was smaller than that for other drugs. Shortages of vaccines emerged before the pandemic.

So too while the number and quality of vaccines improved, Canada's once robust, vaccine-making capacity dwindled. Observers suggest that having originated in academic settings, their business models were highly inefficient. Beginning in 1972, the famous Connaught Laboratories, formerly affiliated with the University of Toronto, was taken over, first by the Canada Development Corporation, a government-owned holding company; then, through various sales and mergers, it belonged to the French company Sanofi-Aventis. Similarly, the commercial branch of Quebec's prestigious Institut Armand-Frappier, which affiliated first with the Université de Montréal and later the Université de Québec, was sold to the British company GlaxoSmithKline. While some facilities of these foreign-owned entities remain in Canada, their commercial production lines were severely restricted.

Each new infectious outbreak spawns more research to develop an effective vaccine. SARS inspired several efforts that were unsuccessful and then abandoned when the virus vanished. A similar response occurred with MERS. Ebola also resulted in many attempts and numerous trials, especially following the 2014 outbreak. To date, however, only one Ebola vaccine has been licensed by the US FDA. It was to Merck, in December 2019; scientists with the Public Health Agency of Canada had been participating in its development since 2003. Other Ebola vaccines are in the pipeline. One was approved by China in 2017; it had been developed by CanSino, a biotech firm, founded in 2009. Once again, scientists at Canada's National Research Council had collaborated in the project.

After testing in animals, every new vaccine (and for that matter all pharmaceutical products) must go through three phases of clinical trials in humans: phase one – a small group to see if it is safe; phase two – a larger group stratified by age and gender to check for doses and side effects; phase three – a huge group to assess its effectiveness compared to placebo (nothing) or to other agents (if any exist). To be effective and prompt, a phase-three vaccine trial must be conducted where the disease is rampant, otherwise it becomes nearly impossible to measure any benefit or harm between those vaccinated and those not. In the case of vaccines (as opposed to drugs), it is important to measure the antibody response in the blood of trial subjects. Each country then analyzes the scientific data to make sure that the product is safe and effective. Trials are expensive, and they need the capital of rich pharmaceutical companies. The WHO also runs a vaccine approval process that aims to give recommendations and advice, especially to countries that cannot afford the rigorous evaluation process; however, some firms ignore WHO or treat it in a cavalier manner, scarcely supplying it with any data.

By early 2020, it was said that, on average, ten to fifteen years was required to make any new vaccine, and no vaccine had yet been made against a coronavirus. Nevertheless, the world had become accustomed to the promise and power of vaccines in handling infectious disease. Many universities and commercial laboratories rose to the pandemic challenge, turning the sleepy vaccine sector into a pressure cooker of biotech prowess, fueled by nationalistic aspirations and urgent demand, and blending academic wisdom with commercial know-how and manufacturing capacity. Shortages were predicted even before the vaccines were complete.

THE EARLY VACCINES

The COVID-19 vaccine work became a remarkable scientific success story. But at the time of writing, in early 2022, it is also a terrain of soaring anxiety. As governments poured money into the effort, all the steps were sped up: research, trials, and production. Originally projected to be at least eighteen months or even two years into the future, the first vaccines were released in less than a year.

As mentioned in chapter 3, the WHO had joined with other non-profit agencies to launch ACT (Access to COVID-19 Tools), a three-pillar approach to providing equitable access to diagnostics, treatment, and vaccines. The vaccine pillar is called COVAX, a commitment to helping countries obtain vaccines through funding and shared distribution. It was launched in April 2020.

Owing to the urgent situation, scientists mobilized all the known methods of making vaccines. To save time, developers decided to allow the classic trial phases one, two, and three to overlap with each other. The Trump administration's Operation Warp Speed, a public-private partnership, featured a massive infusion of government funds from the CARES Act, amounting to ten billion dollars for research and development (chapter 2). Between May and August 2020, eight companies became beneficiaries; most were large, well-established firms. Eager announcements of vaccine research would appear online and in social media before the detailed scientific publications could be peer-reviewed, published, and analyzed. Health-care institutions and media outlets set up interactive trackers to follow the progress of vaccine candidates. Some experts worried that the new vaccines were not properly assessed before being applied to mass immunization.

The earliest COVID-19 vaccine to be used on humans was the Russian Sputnik V, released in August 2020. A *viral-vector vaccine* from the

Gamaleya company, it relies on the common and harmless adenovirus to infect host cells, delivering the gene that codes for the coronavirus spike protein. The viral gene then marshals the host-cell RNA to make the spike protein that will trigger the body's immune response against it. Adenoviruses are safe. They have already been used in genetic engineering for some time and this viral-vector method is the basis of the successful Ebola vaccines. Russia announced that it started injecting the vaccine in its most vulnerable citizens on 11 August 2020. The decision was criticized because it was based on results only from phase-one and phase-two trials with no phase-three information; the makers promised that phase-three data would come from results in the vaccinated vulnerable citizens. Vladimir Putin touted the early roll out as a personal victory, while scientists around the world worried about the risks of ignoring phase-three trials. Cartoonists portrayed Putin shirtless (as usual) planting a syringe and hammer-and-sickle flag on the pocked, moonscape surface of the virus. In the following months, Sputnik V was distributed to approving, allied customers in Belarus, Argentina, Guinea (experimental use), Bolivia, Algeria, Palestine, Venezuela, Paraguay, Turkmenistan, Hungary, UAE, Serbia, and Iran.

Less than two weeks later, China followed with at least two vaccines, Sinovac and Sinopharm, both of which use an *inactivated virus*, which is a weakened form of actual SARS-COV-2. This technique is the old, tried-and-true method, dating back to the nineteenth century and Louis Pasteur's work on rabies. The shot contains bits and pieces of killed virus, which will trigger the immune response but are incapable of causing illness. In October, China announced results in the *Lancet* from phase-one and phase-two testing of a third vaccine, based on the same technique, which it began using on students in November 2020. Again, emergency use would constitute (or substitute) for phase-three trials, which had to be conducted elsewhere because China had virtually eliminated COVID-19 disease from within its borders. By early 2021, Chinese-made vaccines were marketed in Asia, South America, the Middle East, North Africa, Serbia, and Turkey. But in early January 2021, Brazil reported that Sinovac gave disappointing results at only 50 per cent efficiency, which meant that vaccinated people were only half as likely as the unvaccinated to develop infection when exposed to the virus.

In October 2020, Russia announced the safety of another vaccine, EpiVaxCorona, a *peptide vaccine* that uses synthetically constructed proteins from SARS-COV-2 attached to a carrier protein and an aluminum-containing adjuvant. This technique is relatively new, but it has

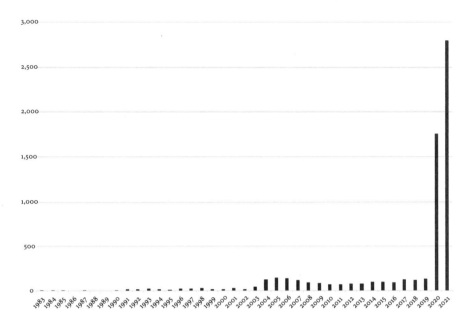

Figure 8.1 | Number of articles per year on coronavirus spike protein, 1983 to 2021 (Source: Topic Keyword Search, Web of Science, 30 January 2022.)

been used for several years in oncology to trigger immune responses against the antigens of cancer cells. So far, it is fully approved only in Turkmenistan, with emergency use in a handful of other countries, although results were presented to the WHO in March 2021.

In the West, tremendous fanfare greeted the arrival of the Pfizer-BioNTech two-dose vaccine, approved in the UK on 2 December 2020, less than a month after its successful trials were announced. Being the world's first *m-RNA vaccine*, it was heralded as a triumph of scientific ingenuity. Some critics were suspicious of the new type of vaccine because it had appeared quickly and trials were small; however, few realized that it had emerged from more than thirty years of research on coronaviruses. In fact, both SARS in 2003 and MERS in 2014 had prompted molecular biology work on coronavirus spike proteins with a view to producing future vaccines, although no vaccine had yet appeared.

The temporary effect of those two previous, coronavirus outbreaks on scientific publishing can be seen the undulating rise and fall of articles published yearly in figure 8.1. The advent of COVID-19 unleashed a flurry of activity that made the previous decade of work seem to pale in comparison. The blinding speed of the mRNA vaccine owed a debt to many years of prior research.

BioNTech was a little-known biotechnology firm, situated in Mainz, Germany, founded by German scientists Ugur Sahin and his wife Özlem Türeci, both with roots in Turkey. They had been working on cancer treatments with monoclonal antibodies and messenger RNA (mRNA), and they had partnered with Pfizer in 2018 for making influenza vaccine. With funding from the Bill & Melinda Gates Foundation, they had also turned to producing treatments for tuberculosis and HIV. The mechanism of the mRNA vaccine is to deliver a piece of the novel coronavirus's own nucleic acid (RNA) inside a lipid nanoparticle into the cytoplasm of human cells, which will then begin making the spike protein, which then stimulates the immune response. Stock in their company soared, making the couple billionaires and their modest lifestyle (they don't own a car), the object of curiosity. Germany celebrated their achievement as a manifestation of successful immigration and public-private partnership. Media outlets around the world began predicting a Nobel Prize for 2021. (It didn't happen.)

Countries clamoured to approve the new mRNA vaccine; many, like Canada, had already committed to purchasing millions of doses before the trials were complete. Thirty-four nations in every region of the globe had approved it for emergency use by early 2021, a number that swelled to more than 120 by the end of the year with ten other nations giving it full approval, including Australia and Canada. But at first, the two-dose vaccine posed great difficulties for transport and storage. It had to be kept at the extremely cold temperature of minus 70C or it would lose its effectiveness. Few places had the infrastructure to maintain those conditions, especially countries like Canada, where distribution involved vast distances. A scramble to locate, obtain, and install specialized freezers then ensued.

Just over two weeks later, a second mRNA two-dose vaccine was approved by the European Union. It was made by the American-based firm, Moderna, which had been a beneficiary of Operation Warp Speed funds. While it too used the mRNA-and-nanoparticle technology, like the Pfizer-BioNTech vaccine, it did not require extreme cold for storage, making it easier to transport. By early 2021, it had been approved for emergency use in eleven nations, including Canada, and by year's end, the number had grown to a hundred, with four countries, including Australia and Canada, giving it full approval.

The Moderna company had been founded in 2010 by three scientists, including Canadian stem-cell scientist, Derrick Rossi. Its explicit purpose was to develop drugs and vaccines with modified RNA (hence Mod-e-RNA). To modify RNA, they had licensed the brilliant chemical achievement of Hungarian-born Katalin Karikó and her University of Pennsylvania colleague Drew Weissman. In 2005, Karikó's team had resolved a major obstacle in mRNA research, which had been hampered by causing a severe immune response (like a cytokine storm) when injected into animals. Essentially, they located and removed the offending part of the molecule that had triggered the dangerous response. This same technique was also adopted by the founders of BioNTech. No longer with Moderna, Rossi believes that Karakó and Weissman deserve the Nobel Prize in Chemistry.

By the beginning of the pandemic, Moderna had done a brilliant job of fundraising, but it had yet to release a single finished product. Nevertheless, it became the first firm to take a COVID-19 vaccine to human trials, reaching phase three in late July 2020 and completing the development work in a near tie with Pfizer-BioNTech. In fact, the two companies had raced neck and neck to meet the avid demands of an exhausted global population, although their styles were vastly different. Moderna used the business model of a pharmaceutical company with extensive capital, vigorous fundraising, and trade secrets. It was projected to earn US$18 billion in 2021; however, it got into dispute over patent rights, maintaining that federal scientists had played a lesser role than was claimed and despite the generous government support. Moderna's legal action meant that its vaccine could not be disseminated and shared; a flood of negative publicity resulted and it eventually backed down. In contrast, BioNTech operated more along the lines of an academic entity, releasing its discoveries in multiple publications; Sahin has more than six hundred articles to his name. BioNTech's American partner, Pfizer, declined Operation Warp Speed funding to keep its scientists free of obligations; it had promised to absorb all costs should the venture fail.

Another long-awaited, *virus-vector vaccine* finally came out of the multinational AstraZeneca company, headquartered in the UK and partnered with Oxford University. Like Moderna, it had taken Operation Warp Speed funding, and its two-dose vaccine was first approved in the UK by 29 December 2020. Although it was widely used with apparent safety and effectiveness in the United Kingdom, where it had been invented, this vaccine would run into multiple difficulties, including possible blood clotting in recipients; some of these concerns seemed to be generated by nationalistic agendas (chapter 11).

Days later, on 3 January 2021, India approved its own inactivated virus vaccine, called Covaxin, developed with government support by Bharat Biotech, which already had numerous vaccines in its portfolio. The tremendous capacity to scale up production in India to as many as 500 million doses per month is countered by a skeptical population. In its first month, only about two million people accepted vaccination. So far, it has full approval only in India, emergency use in some African and Latin American countries, but it is widely accepted for travel.

By early February 2021, less than a year after the declaration of the pandemic, a total of nine vaccines had been approved for commercial release in at least one jurisdiction. COVID-19 was raging in so many places that finding suitable locations for phase-three testing was not difficult. Perversely the great extent of the pandemic facilitated vaccine development.

The Johnson and Johnson vaccine (also called Janssen vaccine) was approved for emergency use in the United States on 27 February and by Canada on 5 March 2021. Its mechanism of action also relies on a virus vector, like AstraZeneca and Sputnik V; however, unlike all other vaccines described so far, it required only a single dose. Some were touting its usefulness for people who would be difficult to follow for giving timely second doses, such as migrants, refugees, and the homeless.

Another vaccine candidate, which relies also on viral vectors, found early emergency use approval: Cansino out of Wuhan, China. It had been used on the Chinese military even before Russia began administering its Sputnik V, and before any late-stage trials had been conducted. It has been fully approved only in China, although nine other countries have granted emergency use authorization.

Another vaccine, from Novavax, has a novel approach. It combines the protein (peptide) vaccine with a nanoparticle delivery system. The Maryland-based biotech company, Novavax, was the smallest of eight entities to receive Operation Warp Speed funding. Coming a bit later in the pandemic story, its trials were able to assess its activity against new SARS-COV-2 variants from South Africa and the UK; results in summer 2021 were encouraging. Phase-three results were not available until October. Plans were afoot for Novavax to produce its vaccine in a new facility in Montreal, but production was delayed into 2022. In fall 2021, it finally applied to WHO for emergency use approval, and was granted full approval in two countries in early 2022.

At least fifty-five other vaccine candidates have been in the pipeline, most using variations on the technologies described above. Three have already been abandoned; others were delayed. Reflecting the novelty in the cutting-edge of biotechnology, seventeen (or more) are products of collaboration between university-based scientists and the private sector, often with government support. Remarkably, Cuba has produced and begun using two home-grown vaccines: Soberana, and Abdala, which both use an inactive portion of the virus. Neither has yet received approval from WHO, but the country, so long subjected to trade embargos, aimed at full vaccination of its population by the end of 2021 and began vaccinating children as young as two years old. By January 2022, 85 per cent of Cubans were fully vaccinated.

Canada has had at least three vaccine candidates in development with financing coming from federal government and commitments to purchase: Quebec-based Medicago; Calgary-based Providence Therapeutics; and CanSino in collaboration with China. The first contribution of almost C$200 million came in late March 2020. The CanSino vaccine, described above, was supposed to run clinical trials together with Dalhousie University, but the arrangement fell apart in summer 2020 when substrate was blocked at Beijing airport by Chinese customs. Many attributed the delay to continuing strained relations between China and Canada: the latter's arrest of Huawei executive Meng Wanzhou to honour an American extradition request; and China's arbitrary detention of Canadian businessmen Michael Kovrig and Michael Spavor. All three were finally released on 24 September 2021.

Lacking any development capacity, Canada also began reserving millions of doses of vaccine in August 2020 in signed agreements with seven companies: Pfizer-BioNTech, Moderna, AstraZeneca, Johnson and Johnson, Novavax, Sanofi, GlaxoSmithKline, and Medicago. The process entailed the government paying some money up-front for development with commitments to pay the remainder upon approval and receipt of the products. Canada also invested in expanding existing facilities and partnering with pharmaceutical firms to build new factories, both long-term propositions aimed at rebuilding domestic production.

Canada's plans were criticized, on one hand, for coming too late and on the other, for selfishly reserving so much vaccine, enough to immunize the entire population four times over. The extreme abundance in orders was owing to skepticism over how quickly or safely any company could invent an effective vaccine, convey it through the approval process, and ramp up production to meet orders. Nevertheless, Canada was one of the first developed nations to join the WHO's COVAX initiative, and it promised to share excess vaccine with other nations through COVAX. Initially Canada donated US$250 million to COVAX, an amount bested only by the UK and approximately double initial donations from the Gates Foundation, France, Germany, and Japan. The United States refused to join until after Joe Biden came to office; its first donation appeared in February 2021. What many people did not realize is that contributors to the COVAX stockpile could also avail themselves of vaccine.

Despite Canada's positive record on COVAX, its attitude to its own access-to-medicines policy has come under scrutiny. In early March 2021, the small Canadian biotech company, Biolyse, was prepared to make vaccine and appealed to Johnson and Johnson for a voluntary licence to manufacture its vaccine for Bolivia, which had indicated that it would be a willing buyer. The request was rejected. The next option was to apply for a compulsory licence under the Canadian Assess to Medicines Regime, an exemption (waiver) to the Agreement on Trade-Related Aspects of Intellectual Property Rights (TRIPS) for patent protection, which only Canada, alone in all the world, has ever exercised and that only once in 2008–09. The quantity of red tape that Biolyse encountered in trying to meet requirements prompted Muhammed Zaheer Abbas, a researcher at Queensland University of Technology, to conduct a deeper investigation, published by the South Centre in September 2021. Abbas concluded that the Trudeau government is "not politically inclined to the humanitarian needs of poorer countries and has been trying to hide behind the legislation," using "unjustified formalities" and revealing "contradictions" between its statements and its actions.

The early triumph of vaccines in the COVID-19 pandemic was quickly tarnished by the hostile, nationalistic scrambling that would soon follow in obtaining, administering, and encouraging acceptance (chapter 11). And it was humbled by the inevitable arrival of variants that seemed impervious to their power (chapter 12).

PART THREE

The Following Waves, and Beyond

"Wave after wave, each mightier than the last."

Alfred Lord Tennyson,
Idylls of the King

9

Analyzing the Spread:
Statistics, Damn Statistics, and Lies

Rather than track the global spread through time as seen in part one, this chapter will select a few examples of later spread, using statistics to illustrate the discussion. Numbers are abundant. From the pandemic's inception, national and international health organizations, media outlets, and private citizens began running interactive online trackers (chapter 5). The course of the pandemic became riddled with statistics. Mark Twain, whose manipulated quip adorns this chapter's title, classified statistics as the worst of lies. The lies may not be deliberate, but sometimes statistics hide important truths: as baseball's Toby Harrah once said, "statistics are like bikinis, they show a lot but not everything."

Second-wave infections, rising and waning, are anticipated in epidemics. They were seen in fourteenth-century plague, in 1918 influenza, and in recent Ebola. The many causes include environment, climate, living conditions, health vulnerabilities, and human behaviour. Only some of these conditions can be modified. The influenza of a century ago, with its many similarities to COVID-19, saw far worse case numbers and deaths in its second wave than in the first.

LESSONS FROM THE FIRST WAVE

Everyone knew a second wave of COVID-19 would come. Nevertheless, the relief and joy that greeted the end of the first wave and the lifting of lockdowns made it difficult to accept more restrictions when they became necessary again.

Some countries, like Belarus, never really escaped their first wave: numbers just kept oscillating, never falling to low levels. Others, like France and Italy, are already deep into their fourth and even fifth waves

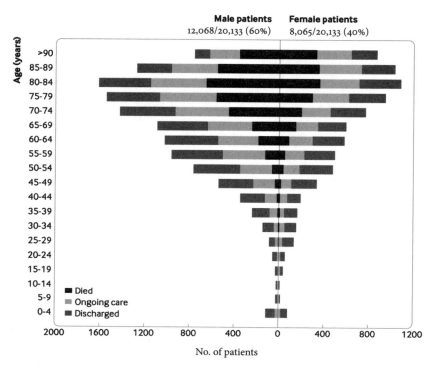

Male patients
12,068/20,133 (60%)

Female patients
8,065/20,133 (40%)

Figure 9.1 | Features of UK patients with COVID-19 as of May 2020 (Source: BMJ, https://doi.org/10.1136/bmj.m1985, 22 May 2020. By permission and with thanks to Dr Callum Semple.)

at the time of writing. The repeated closings and openings are depressing, and they feed into the denial and conspiracy theories. Once again, politicians hesitated to implement the stringent lockdowns that healthcare workers and scientists were demanding, fearing backlash from voters, especially those whose livelihoods were imperiled.

By the end of the first wave, uncertainty over the clinical features of the illness had been clarified. Doctors learned that many other symptoms could characterize COVID-19 than the early experience had suggested. They became aware that a certain proportion of people, called long-haulers, had cleared the virus but continued to feel sick with headaches, fatigue, neurological problems, difficulty concentrating, and what they called brain fog. In 10 per cent or more of patients, these chronic problems persisted for months. Pregnant women were more likely to suffer severe illness or premature labour, while young children were less likely to catch or spread the virus and rarely suffered severe disease. On rare occasions, however, the cytokine storm could affect children too, provoking inflammation of the heart and other organs, called Kawasaki's disease or pediatric inflammatory multisystem syndrome (PIMS). By early February 2021, enough data had accumulated to allow British experts to suggest that the mysterious syndrome affected approximately one in every five thousand children with COVID-19, most of whom had previously been healthy. Genetics also seemed to play a role: 75 per cent of the affected children were Black or Asian. Poverty may have been a factor too.

The identity of people – or demographics – that characterized each wave was not always stable. In May 2020 a report in the BMJ, summarized outcomes by age and gender, based on data from more than twenty thousand patients treated in two hundred British hospitals (figure 9.1). It confirmed greater danger for the elderly and slightly worse outcomes for men than women; hospitalization and long-hauler effects could occur in any age group. The authors kept up the analysis, forming a team called ISARIC₄C Coronavirus Clinical Characteristics Consortium. The team is a UK network within the International Severe Acute Respiratory Infection Consortium (ISARIC), which was created following the 2009 influenza pandemic. It focuses on patient characteristics between and during outbreaks aiming to improve care and inform public health. ISARIC$_4$C had analyzed MERS, Ebola, and Zika outbreaks. It maintains live reporting at its website, where dynamic charts like figure 9.1 can be found with data on an ever-increasing number of patients. Over time, cases shifted to younger people, but deaths still predominate among the elderly.

In trying to flatten the curve, doctors had also learned, through trials, how to properly wield various treatments, although none were cures (chapter 7). Oxygen was better when given by a mask rather than a tube. Dexamethasone and remdesivir could help the seriously ill in intensive care but were of little use in mild illness. Anticoagulants might salvage those with an impending cytokine storm. Colchicine might reduce disease progression at the outset. Ivermectin was refuted as effective treatment, only to enhance its cachet. Fluvoxamine, long used for mental illness, was finally endorsed for COVID-19 in Canada in January 2022. Case fatality rates, originally at 3 per cent or more, improved as caregivers learned helpful strategies for treating active infections; however, when health systems were overwhelmed or nonexistent, case fatality rates could worsen (table 9.1).

Table 9.1 | COVID-19 in thirty-six countries: cases, deaths, and tests per capita with case fatality rates, March 2021 and January 2022

Country	Cases per 1,000	Deaths per 1,000	Testing per 1,000	Case fatality rate as of Mar. 2021 (%)	Case fatality rate as of Jan. 2022 (%)
Americas					
Canada	68.6	0.8	1,410	2.43	1.18
United States	189.4	2.6	2,528	1.82	1.37
Mexico	31.6	2.3	97	9.0	7.28
Haiti	2.3	0.1	11	1.2	3.04
Argentina	142.6	2.6	647	2.43	1.82
Brazil	105.3	2.9	297	2.5	2.75
Chile	94.9	2.0	1,446	2.4	2.11
Europe					
Belgium	196.0	2.4	2,404	2.7	1.22
Czechia	236.4	3.0	4,436	1.7	1.44
Finland	61.0	0.3	1,607	1.1	0.49
France	192.0	1.9	2,882	2.2	0.99
Germany	90.6	1.4	1,064	2.8	1.55
Italy	128.9	2.3	2,489	3.11	1.78
Portugal	166.8	1.9	2,737	2.1	1.14
Russia	73.2	2.2	1,670	2.14	3.01
Spain	162.3	1.9	1,415	2.3	1.17
Sweden	145.9	2.0	1,497	1.8	1.03
UK	215.3	2.2	6,217	2.9	1.02

Table 9.1 | *continued*

Country	Cases per 1000	Deaths per 1000	Testing per 1000	Case fatality rate as of Mar 2021 (%)	Case fatality rate as of Jan 2022 (%)
Oceania					
Australia	43.9	0.1	2,218	3.11	0.23
New Zealand	2.9	0.01	1,141	1	0.34
Asia					
Bhutan	3.6	0.004	1,697	0.1	0.11
China	0.1	0.003	111	5.15	4.17
Hong Kong	1.7	0.028	3,962	1.8	1.64
India	25.7	0.346	495	1.4	1.34
Iran	72.5	1.541	501	3.42	2.12
Japan	14.1	0.146	237	1.93	1.04
Myanmar	9.7	0.351	112	2.3	3.62
South Korea	13.1	0.119	308	1.7	0.91
Vietnam	19.6	0.353	769	1.4	1.80
Africa					
Egypt	3.8	0.21	35	5.9	5.50
Kenya	5.6	0.10	56	1.7	1.75
Nigeria	1.2	0.01	18	1.3	1.17
South Africa	58.5	1.50	358	3.4	2.56
Uganda	3.2	0.07	47	0.8	2.19
Zambia	15.0	0.20	160	1.4	1.33
Zimbabwe	14.7	0.34	125	4.1	2.32

Source: Worldometer, aggregate data, https://www.worldometers.info/coronavirus/, 11 January 2022.

Global Situation

298,915,721
confirmed cases

| Jan 1 | Apr 1 | Jul 1 | Oct 1 |

5,469,303
deaths

| Jan 1 | Apr 1 | Jul 1 | Oct 1 |

Source: World Health Organization
Data may be incomplete for the current day or week.

Figure 9.2 | Global confirmed seven-day average of new cases and deaths, January 2020 to January 2022 (Source: World Health Organization, https://covid19.who.int, 10 January 2022.)

Protection had also been refined. Resources, such as masks, hand sanitizer, PPE, and tests, were no longer scarce. In some places, they had become overabundant, annoying those companies that had shifted activity to furnishing the increased demand and now were left without clients.

GLOBAL WAVES

The waves were happening at different times, in different places, and later caseloads often exceeded first-wave levels (figure 9.2). Public-health experts kept warning that the threat would continue undiminished

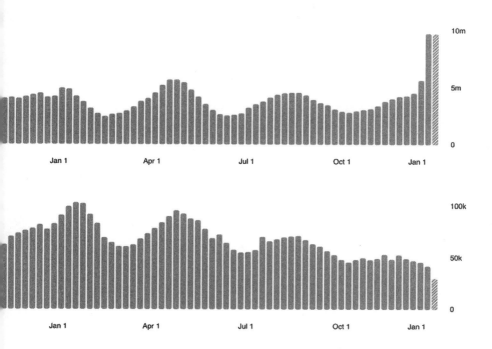

because the virus was new and the entire world was immunologically naive, meaning that no one had natural resistance. The shocking numbers of the first wave were dwarfed by what came next: a second wave in July to August 2020; an even greater third wave from October 2020 throughout the winter. As numbers of new cases dropped, recorded deaths would begin to decline two weeks later. Decrease in global numbers owed most to controls in Europe and North America, while the disease in low- and middle-income countries persisted at an apparently slower but steady rate, hinting at inaccurate counting or massive waves, or both, still to come. The third-wave decline was short lived. By early

March 2021, even as the vaccines were arriving, global cases began to creep upward, and fourth, fifth, and sixth waves lay ahead. Authorities everywhere then had to cope with pushback from people who had grown tired of the social, financial, and psychological hardships that accompany lockdowns (chapters 10 and 12).

Some national leaders also opposed sanitary measures. In March 2021, an alarming spike in disease took place in Brazil, still mired in its unrelenting second wave. Hospitals were overwhelmed. President Jair Bolsonaro appointed his fourth health minister in a year and appealed to the Supreme Court to overturn lockdowns that various regions and municipalities had enacted in the absence of national guidance. Brazil was said to be on the verge of collapse; its caseload approached eleven million and its death toll climbed to more than a quarter million, second only to the United States. In October 2021, a leaked Brazilian congressional document asserted that Bolsonaro should be charged with crimes against humanity for his failure to manage the pandemic.

THE VARIANTS

As anticipated, the virus mutated, although the earliest changes were of little significance. To avoid geographic stigmatizing habits of the past, scientists piously tried to insist upon the anodyne but complicated scientific labeling system. To appreciate the myriad number and complexity of variants, simply conduct a Google search with the words "COVID-19 phylogenetic tree"; you will find bewildering arrays of intertwining mutations. The so-called UK variant, named B.1.1.7, was first detected in September 2020. Ironically, in the UK it was called the Kent variant for the county of origin. The South African variant, called B.1.351, with similarities to the UK variant, was reported in early October 2020. A Nigerian variant, called B.1.525 emerged in late December. The Brazilian variant, called P.1, was first isolated from four Brazilian air passengers who arrived in Japan on 2 January 2021; it differs more from the others. On 14 February, Brazil reported its first two cases of the UK variant on samples taken 31 December 2020, by which time the UK variant had been found in seventy countries worldwide. In mid-January 2021, *Nature* quoted a biostatistician who, like the child in *The Emperor's New Clothes*, stated the obvious: viral naming conventions had become "a bloody mess." Media and the public relentlessly continued to use the easier (but stigmatizing) geographic names.

Preliminary reports suggested that some new variants were said to be of concern, that is, more contagious, provoking more cases and deaths. At first, however, the clinical course of each new variant seemed no worse than the so-called wild type. But the California outbreak in late 2020 was thought to result from a more aggressive variant. Whether or not illness provoked by variants would respond to the same treatments, or be prevented by the forthcoming vaccines, was unknown. By mid-March 2021, increased experience with variants revealed that the UK variant was not only more contagious, it was also at least 55 per cent more lethal and was rapidly becoming dominant in Europe and elsewhere.

Early observations suggested that the forthcoming vaccines would protect against variants, but it was unreasonable and naive to expect that mutations would never be resistant. The relentless assault of new variants meant that crushing the pandemic became even more urgent; the virus should not be given time and opportunity to generate more aggressive strains.

Public-health officials doubled down on their basic recommendations. Flights were blocked from the origin nations, although all three of the earliest variants of concern had already escaped to multiple other countries before the bans. When the highly contagious and more aggressive Indian variant appeared, WHO had had enough of the persistent stigmatizing in geographic naming. On 31 May 2021, it launched a new system of nomenclature by Greek letters. The UK variant became *alpha*, South Africa's became, *beta*, the Brazilian became *gamma*, and the concerning new variant from India was *delta*. Many more variants appeared, marching through the Greek alphabet.

ANALYZING THE WAVES

Epidemiologists use statistics to understand the spread and control of disease. Absolute numbers of cases, deaths, and hospitalizations show the rise and fall of cases. An important tool that allows comparison between countries is the concept of *incidence* – or *rates* per capita – of cases, deaths, and hospitalizations. In a sense, incidence measures how familiar a population has become with the pandemic: the higher the rate, the more likely citizens will know someone who was affected or become affected themselves. Incidence also allows comparison of one region to another, even when populations and densities are very different (table 9.1). Statistics are most reliable in countries that have a high rate per capita of testing – and are skewed downward where the rate of testing is low. Another measure – the case fatality rate – expresses the chance of surviving the disease – how many of the sick will die. It depends on the quality and accessibility of health-care facilities, but it is also affected by other factors, including the general health and average age of a population: older, sicker people do worse. The pandemic was not the same everywhere.

By summer 2020, mathematical analysis of the spread of the disease and its control in various regions of the world was available. One study published in July 2020 by a team from Stanford University suggested that the reproduction number (R_0) for Europe in the early first wave had been greater than four, amplified by travel, public transit, and even walking (chapter 5). Public-health measures that restrict movement and large gatherings could bring the R_0 value down to below one, within an average delay of seventeen days. The authors began predicting what would happen as restrictions were lifted.

COUNTRIES THAT HAD ONLY ONE WAVE (SO FAR)

China, where COVID-19 made its frightening debut, managed to control and almost eradicate the virus. Its late outbreaks were small and quickly contained, by swift action. Over the first two years, the roughly eighty thousand cases and 3,500 deaths of its initial wave crept slowly upwards to just over 117,000 cases and five thousand deaths with small outbreaks in various regions, especially around Beijing by late 2021. All new cases were attributed to returning travellers. Each outbreak triggered an aggressive lockdown. For example, in late October 2021, a hundred cases of *delta* variant spreading across eleven provinces, with thirty-nine in the city of Lanzhou, prompted travel restrictions and a tight lockdown of millions of people lasting several weeks. The aggressive measures typified China's zero-tolerance policy, which it maintained for its 2022 Winter Olympics and beyond with a spring surge in Shanghai.

Aside from China, it is doubtful that any country had only a single wave of COVID-19 in the first two years. Caseloads wax and wane according to season, public-health measures, and behaviours. Also, reporting depends on a country's ability to measure and its willingness to reveal the numbers.

In early 2021, continuous single waves and those just starting to decline were seen in other countries – some with high incidence; some, low. Examples include such variable experiences as India, the Philippines, Ukraine, Poland, New Zealand, and Lebanon where the devastating

Beirut explosion on 4 August 2020 destroyed social, medical, and financial infrastructure. Suffering massive hardship, Lebanon saw its COVID-19 caseload swell with little decline until mid-January 2021, while deaths continued to rise. Its health-care facilities were on the verge of collapse; vaccines were slow to arrive. Lebanon's pandemic plateaued in the summer of 2021 and worsened in the autumn; however, concern over accuracy of its statistics is easily understood. In contrast, New Zealand and Bhutan have had very few cases or deaths in total, but even they – together with most other nations – saw a return of the virus in late 2021 (chapter 12).

INDIA

At first, India seemed to have a single big wave. With more than ten million cases it was second only to the United States in absolute numbers of cases by late 2020; however, its enormous population of 1.4 billion meant that the *incidence* of cases and of deaths was among the lowest of the large economies, where it remains, despite what happened next (table 9.1). By early 2021, cases numbers were declining steadily, well before the Indian-made vaccine had been approved. Scientists scrambled to explain why. A study, released in February 2021 by the New Delhi government, suggested that most of its twenty million citizens already had antibodies against the virus, although they had never been ill and had never been tested! It suggested that a large proportion of the population, at least in this one city, had contracted the virus without knowing it. India – or some of its urban areas – might already be tilting toward herd immunity came the cautiously optimistic but premature conclusion. India was on the verge of a meltdown.

Speculation over herd immunity vanished, as India's brutal second wave began in March 2021 with numbers that far exceeded those seen in its first wave. Within weeks, the case and death rates soared, and the country was crippled by shortages of medications, oxygen, hospital beds, and facilities for decently cremating its dead. Behind this second wave was the new variant, B.1617.2, now called *delta*, more contagious and more virulent than any of its predecessors. It also raised concerns about how effectively the new vaccines would block it, since *delta* did not exist when they were developed. It spread rapidly and became dominant in the UK by May 2021. By late August, Canadians were expecting a *delta*-driven fourth wave. Meanwhile, citizens of India displayed perplexing hesitancy in accepting their own vaccine.

THE UNITED STATES

By January 2021, the United States had suffered a quarter of the world's one hundred million cases. A year later, its caseload had tripled and was still growing. The country displayed an undulating and growing pattern of cases, deaths, and hospitalizations. With its large population, it remains the country with the most cases and the most deaths, 16 per cent of the globe's total, although other countries, especially in Europe, have witnessed a higher incidence in per capita cases (table 9.1).

But this overall national pattern is constructed from widely varying patterns that emerged over time from individual states. New York state, for example, so grievously affected in the first wave, displayed even more cases in the second wave, but with fewer hospitalizations and deaths. The lower second-wave death rate hints at better management of the disease through experience and also, perhaps, to a shift to younger, healthier sufferers, owing to improved protection or sadly to the earlier loss by death of the elderly at risk, or both. Recall it is likely that the actual case count in the first wave was much higher than what had been reported due to lack of testing capacity.

California on the other hand, despite seeing the earliest travel-related cases, was largely spared during the American first wave. But when the disease finally took hold of this populous state, both hospitalizations and deaths exploded with no decline until well into January 2021. Similar patterns were found in Texas and Florida.

The Dakotas, however, saw a dreadful wave in autumn 2020. Its absolute numbers of cases and deaths were smaller than in the more populous states, but the proportion of citizens affected – *incidence* – was much greater. The smaller numbers in sparsely populated North Dakota were proportionately much higher; in fact, they were temporarily the worst in the entire country, far exceeding national averages and approaching the carnage of New York City's first wave.

Vast regional differences, like these, persisted as the United States went on to suffer third, fourth, and fifth waves. The third wave may eventually prove to have been the deadliest, taking place before vaccination was widespread. Where jurisdictions refused or were slow to enact public-health restrictions, COVID-19 was worse.

Politics is always a factor in public health. Furthermore, the pandemic was taking place on a backdrop of the fraught federal election. Many cases resulted from the 6 January 2021 storming of Congress – at least

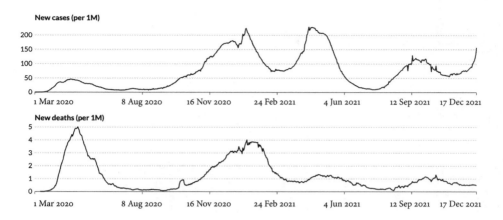

Figure 9.3 | Incidence of COVID-19 cases and deaths in Canada, March 2020 to December 2021 (Source: Our World in Data, https://ourworldindata.org/covid-cases, accessed 26 January 2022.)

thirty-eight police officers tested positive in its wake; some argue that it became an undocumented superspreader event as exposed protestors scattered across the country in its aftermath. It is sobering, but scarcely surprising, to observe that the highest rates of infection appear in states where the stringency index has been the lowest; most often states with Republican governments. These differences emerged as early as May 2020 and persisted well into 2022. Media outlets showed how COVID-19 began in the crowded Democratic states, but by mid-2021, cases and deaths were more numerous and growing in regions voting Republican in the 2020 election.

Later in the pandemic, the Republican governors of Florida, Texas, and Virginia used legislation to oppose public-health recommendations, going so far as to ban mandates for vaccination and masking. Some suggest that they may suffer for these decisions in public opinion and future elections, although they found fans in Canada.

CANADA

Like the United States and the rest of the world, Canada had severe second and third waves with many more cases than in the first wave and many deaths (figure 9.3). Sometimes, caseloads were burdensome, but

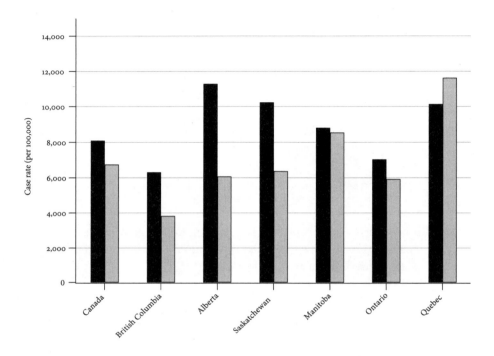

Figure 9.4 | Total cases and deaths per one hundred thousand in Canada, by province and territory (Source: Joshua Lipton-Duffin, based on data from Health Canada, https://health-infobase.canada.ca/covid-19/, 11 January 2022.)

new cases never exceeded ten thousand daily until the *omicron* wave when numbers soared to triple previous maximums (chapter 12). The disease spread unevenly across the country, as it did everywhere else. Regional statistics show several things:

· most cases and deaths occurred in the populous provinces of Ontario and Quebec;
· second- and third-wave caseloads increased over the first, but deaths declined;
· during early waves the Atlantic provinces had a fortunate situation;
· during the second, third, and fourth waves, the West saw increased problems.

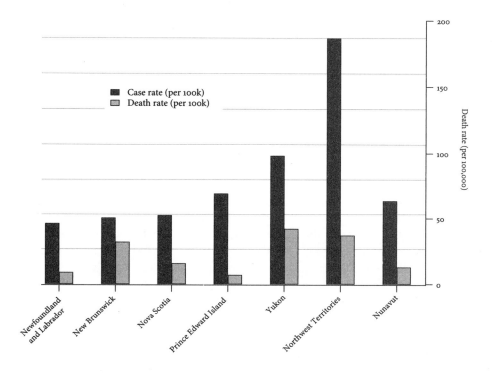

By January 2022, Canada's total incidence of cases was more than 6,600 per one hundred thousand population and total death rate around eighty-one per one hundred thousand. Both statistics were a third of those in the United States, but more than double what had been reported for India to that date and despite that country's massive second wave (table 9.1).

Quebec led the country in per capita cases and deaths because of its severe early waves (figure 9.4). If, instead, the graphs were to portray absolute numbers rather than numbers per capita, figures for the western provinces and the three territories would appear smaller, owing to the larger populations of Quebec and Ontario.

These national figures hide isolated horror stories. Nunavut had suffered no cases at all until 6 November 2020. But soon after, Arviat, a tiny

Inuit community saw 10 per cent of its 2,657 people develop the infection, accounting for 250 of the territory's total of 294 cases by February 2021. The crowded living conditions, the extreme cold, and the numerous places with unsafe drinking water meant that Indigenous communities were vulnerable, placing them at the top of the priority list for vaccination. Fortunately, few deaths were ascribed to these northern outbreaks; however, citizens of the territories point out that they must travel south for complex health care for other diseases; some caught COVID-19 in southern provinces and died there, or they brought infection home with them. They complained that their deaths were not being counted properly.

An early 2021 outbreak occurred in Newfoundland. It, too, had been unscathed until 7 February when eleven cases appeared at a high school in the St John's area, followed by thirty cases the next day, and one hundred soon after. Authorities acted quickly to impose a two-week circuit-breaker lockdown, initially for the region, then for the entire province, with closure of bars, gyms, and schools, work-from-home orders, limits on gatherings, and extension of quarantine. They began looking for evidence of variants and found that the culprit proved to be the more contagious B.1.1.7 UK variant (later called *alpha*). The provincial election scheduled for 13 February had to be halted and converted to mail-in voting only – a first for the country. Anger greeted voting arrangements; people wrestled with downloading ballots and a winter storm meant that deadlines for mailing ballots were repeatedly extended until 25 March. The minister of health, John Haggie, a physician and former president of the Canadian Medical Association, echoed a recurring theme among the harried authorities, when he stated: "The only thing that should be spread is kindness."

By late March 2021, the Newfoundland outbreak was under control and the eastern provinces began discussions about re-creating their Atlantic Bubble (chapter 2). However, persistent cases in Nova Scotia and New Brunswick hampered the plans.

Western Canada suffered severe second and subsequent waves making the three prairie provinces among hardest hit regions of the country. Exhausted health-care professionals became angry and frustrated with their provincial governments for hesitating to invoke restrictions. The mounting cases numbers had filled hospitals beyond capacity.

To the incredulity of many observers and the exasperation of health-care workers, Alberta and Saskatchewan decided to lift all restrictions in July 2021, moving, they said, "from pandemic to endemic." Alberta's Premier Jason Kenney made splashy political announcements about being "Open for Summer." Premier Scott Moe in Saskatchewan followed

Rate of death from COVID-19, by province, since Aug. 1

Lines indicate the number of deaths over the previous 14 days per 100,000 population in each province.

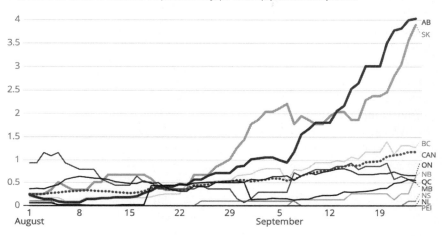

Figure 9.5 | Rate of death in Canadian provinces, August to September 2021 (Source: CBC News, based on Public Health Agency of Canada data, https://www.cbc.ca/news/canada/saskatchewan/sask-covid-ndp-1.6192596, 29 September 2021.)

a similar course but more quietly; however, Saskatchewan soon led the country in per capita cases and deaths. Both provinces appealed for help from the military and the federal government, while other provinces loaned health-care workers and accepted patient transfers. This time, however, Manitoba reversed its previously lax course, which had delivered a challenging third wave, and instead maintained control measures. By autumn 2021, Alberta and Saskatchewan had worse outcomes than Manitoba (figure 9.5). The difference mirrors that of Sweden and its Scandinavian neighbours in the first wave (table 6.1). Other infections, such as influenza and seasonal colds, also declined sharply with the sanitary measures against COVID-19. Public-health measures do make a difference!

The Northwest Territories (NWT) suffered a surge in the fall of 2021. Although the absolute numbers seemed relatively small. When its caseload peaked at more than 450 active cases, the sparsely populated NWT briefly displayed the highest incidence in Canada. Most cases were in the capital region of Yellowknife.

Health Canada tracked cases and deaths from those living on First Nation reservations. The combined total case rate was the highest in the country and the death rate, second only to Quebec. However, the agency admitted that the numbers were probably falsely low for lack of testing in remote places. Like the Arviat outbreak, these statistics reminded the country, once again, of the dreadful living conditions of these isolated communities where people live crowded in inadequate housing with unsafe water supplies.

By late 2021, Alberta had displaced Quebec for the largest caseload per capita; however, Quebec retained its top position for deaths per capita, earned in the devastating first and second waves before the disease was understood and vaccines were available. Two years since Wuhan saw its first cases, most Canadians had become inured to the grim rhythm of waves, grudgingly tolerant of controls, and hopeful that vaccination would bring an end to the tedium. Little did anyone realize that this precarious, oscillating stability was about to be inundated by a wave so great that it would defy measurement (chapter 12).

GLOBAL CONTEXT

As the pandemic wore on, COVID-19 declined with temporary lock-downs and swelled when and where restrictions were lifted too soon. Recall Wuhan remained tightly closed for a long time after its earliest cases waned. With relaxation of controls, public-health officials antici-pated that the disease would spread again; politicians hoped (and some pretended) that it would not. Pressure to lessen the duration and sever-ity of lockdowns came from multiple sources: business owners facing financial collapse; various deluded champions of civil liberty; right-wing rhetoric magnified by social media; pediatricians, stressed teachers, and parents coping with closed schools and the alienation of online classes.

Most countries saw second, third, and fourth waves far worse than the first, with incidence levels that lie between those of Canada and the United States (table 9.1). Some saw fifth and sixth waves. Countries with good health-care facilities learned how to manage the disease, survival improved, and case-fatality rates fell. Differences in fatality rates for March 2021 and January 2022 are shown in table 9.1. Other reasons for improved survival include the effect of vaccines and the demographic extent of the virus as it moved from the elderly to ever younger people

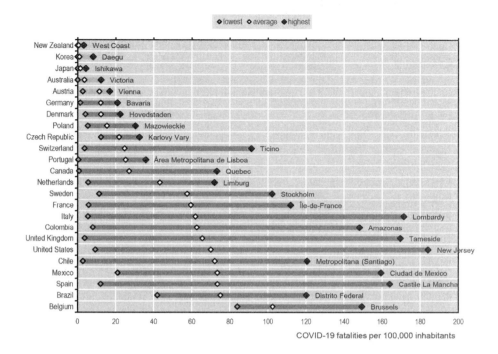

Figure 9.6 | Average death rate and range per capita of and within twenty-four countries (Source: OECD, The Territorial Impact of COVID-19: Managing the Crisis across Levels of Government, 10 November 2020, Fig. 2, p. 6.
By permission and with thanks to Dorothée Allain-Dupré, OECD.)

who were less vulnerable. On the other hand, fatality rates worsened where health-care facilities that had previously been coping were overwhelmed with surges: for example, Argentina, Haiti, Russia, Myanmar, and Vietnam. They also increased with more aggressive variants.

The average death rate per capita for a country, as seen in table 9.1, fails to express what might be wide regional differences within its borders. The graph from the OECD from November 2020, gives a good idea of the considerable range of mortality within each country, as well as the national average. Densely populated urban areas usually suffered most (figure 9.6).

China's overall numbers per capita remained low, but its case fatality rates stayed high because of its initial outbreak; by January 2022 too few new cases had since appeared to reverse that statistic. In the beginning, the absolute lack of testing and the inability to identify mild or asymptomatic cases meant that only the seriously ill were diagnosed, and many died.

Initially managing the pandemic well, Japan was deep into the rise of its fifth wave in late July and early August 2021, when it hosted the delayed Summer Olympics. Only athletes were allowed to attend. Many experts, inside and outside the country, criticized Japan and the International Olympic Committee for forging ahead with the games instead of cancelling them. But Japan maintained strict controls and heavy testing within the athletes' village. Most of those who tested positive were locally engaged Japanese officials. Nevertheless, soon after the games, Japan's fifth wave swelled to become its most severe to date; hospitals were overwhelmed. Whether or not the Olympics had triggered this surge is a matter for debate. On 30 September when the total death toll had reached 17,000, Prime Minister Yoshihide Suga resigned responding to sharp criticism; he was only a year into his term of office.

Across Europe the lockdown measures were highly variable and followed different orders in different jurisdictions; sometimes implemented by regions within a country. They included closing all nonessential businesses and restaurants except for takeout, closing schools, cancelling religious, sports, and cultural gatherings; limiting the numbers who could gather privately; curfews beginning at ever-earlier hours; and closures of external and internal borders, except for vital supplies.

The measures were tempered by protests, where people felt free enough to object and governments were willing to heed the complaints. Stinging from the *gilet-jaune* manifestations of 2018 and 2019, France tried to avoid a third lockdown in spring 2021 with some bizarre recommendations: bars, restaurants, and shopping centres would close, curfews and outdoor masking continued; yet schools would stay open, together with a selective array of businesses: hair salons, bookstores, and chocolate shops. The leaders were strongly criticized by their own public-health experts, while the tepid, ever-changing rules provoked anger, confusion, and despair. French restrictions proceeded city by city, as the variants spread. Meanwhile, with a similar caseload, Italy invoked another national lockdown over much of its territory from March into early April 2021.

High case rates were seen in the UK, as well as France, Belgium, the Czech Republic, Portugal, and Spain (table 9.1). Finland continued to contrast markedly with Sweden, proclaiming the value of its stringent public-health measures. But the credibility of comparative statistics was constantly called into question. Some contended that, at first, Belgium had been counting every death as a COVID-19 death, even when no tests had been (or could be) performed, making its numbers seem worse than the reality.

Unlike Belgium, both Russia and Mexico were under-reporting. Long suspected of releasing inaccurate figures, Russia finally admitted in late December 2020, that it had been undercounting COVID-19 deaths by a factor of three; in other words, its deaths and case-fatality rates were falsely low. But the increased numbers were slow to appear in the official statistics. The problem lay in a lack of testing and hesitant diagnosis: deaths were not counted as COVID-19 unless an autopsy had confirmed it. In October 2021, Russia acknowledged a surge in cases and deaths and instituted a nation-wide lockdown, its first since the early pandemic.

For its part, Mexico continued to have problems with testing. The incidence of cases and deaths seemed low, but its case fatality rate was high at 9 per cent in March 2021, declining over the following year to 5 per cent (table 9.1). With Peru, Mexico's fatality rate is surpassed only by impoverished, war-torn Yemen where case numbers are meaningless, and a fifth to a third of the handful diagnosed with the disease have died. These terrible statistics likely reflect an absence of testing: the disease is recognized only when patients fall seriously ill.

By early fall 2021, many British citizens, fed up with the rules, abandoned them. Notwithstanding its strong vaccine uptake, British cases soared when compared to other countries in Europe and North America with similarly high vaccination rates, even when compared to India. And its death numbers were among the highest in Europe. Travel restrictions against the UK were proposed, especially by France wrangling with a Brexit-triggered dispute about fishing rights. Soon after, Germany's case numbers soared.

When the Japanese prime minister resigned over the seventeen thousand COVID-19 deaths in his country, pundits pointed out that the death toll in the UK, with just half the population of Japan, was now approaching 137,000 – a per capita figure sixteen times worse. Why did British politicians not likewise accept responsibility, and resign?

Deaths are harder to ignore than cases. Where testing is low, case fatality rates can appear high (e.g., Mexico). More testing would identify asymptomatic cases, enhance controls, and tend to reduce spread and improve fatality rates. Conversely, it is difficult to know what to make of favourable rates of cases and deaths where testing is low (e.g., Egypt, Vietnam). But in Vietnam active contact tracing had been applied well before tests were available – at least at first; it later lost control in summer

2021 with a four-day holiday and arrival of the *delta* variant. Chile's raw totals of cases and deaths were remarkably similar to Canada's; however, its population is roughly half, meaning that the mortality per capita was twice as high. All these figures will no doubt have changed since they were gathered in early 2022. Readers can easily update the statistics at many websites, including the WHO, Our World in Data, Global CoVID-19 Tracker, and Johns Hopkins University COVID-19 Dashboard.

Seeking to explain why low-income African nations seem better off, observers have suggested that it could be an illusion, as scant testing hides actual infection numbers. If African nations truly have less COVID-19, then reasons may lie in the same possible advantages considered in the first wave – relative age, respected public health, favourable climate, genetics, and so on (chapter 3). Did the statistics also reflect better natural resistance among various peoples? A study of random postmortem testing from Lusaka, Zambia, suggested that fully 19 per cent of the corpses of people who had died of non-COVID conditions without being hospitalized were positive for COVID-19. The results could reflect a lack of testing. But this study also raises the possibility that, perhaps, among the many variants of coronavirus are milder forms that might spread herd immunity without provoking illness, like that hinted at by the early study on New Delhi cited above. As Canadian infectious disease physician Andrew Morris wrote on 9 February 2021: "Virus is happy (it gets to replicate), and people are happy (they don't get sick)." Acknowledging the need to rid the world of all forms of the virus, aggressive and mild, he continued, "They need vaccine just like we do."

Of course, the numbers are only as good as their providers. Suspicions were rife that some nations were lying to their citizens and the rest of the world. In certain places, lack of transparency in all things is ingrained and has even been measured by the Corruption Perceptions Index that indicates how confidently we can rely on the statistics. Countries willing to be honest, but with few resources, may simply not possess the ability to test, trace, and report. Furthermore, running in parallel with the statistics and the hard work of health-care providers (and tending to counteract all their efforts) was that parallel epidemic – the infodemic – of conspiracy theories, misinformation, disinformation, and lies, the full extent of which was difficult to comprehend (chapter 10). These falsehoods were not aimed at a disaffected fringe; they had backers among powerful leaders, such as Donald Trump, Jair Bolsonaro, and the fraudulently elected dictator of Belarus, Aleksander Lukashenko, who contended that belief in the existence of COVID-19 was a psychosis.

10

Beyond the Numbers

In this chapter, we examine the human costs of the pandemic: pain of disease and death; hardship of controls; extra burden on poor nations; impact on elections, and political tinkering with policies, such as the health-versus-wealth debate, all within the context of other disasters, which seemed to arise with extraordinary frequency during 2020 and 2021. We also examine the nefarious workings of the parallel infodemic that fed on the fears of unhappy people.

HUMAN COST

The statistics of cases, deaths, and hospitalizations hide heartbreaking personal tragedies that were multiplied over and over around the globe. Every one of the millions who died of COVID-19 had suffered a painful end and left loved ones in sorrow. Family members were unable to visit the sick or attend the dying. Nurses helped patients exchange final words with families through telephones and digital devices. To prevent and cope with outbreaks, care homes adopted public-health restrictions, forbidding in-person visits, even for residents who remained healthy; sometimes, restrictions lasted for months on end. The daily news brought poignant scenes of forlorn people waving at windows, yearning to connect, wishing they could hear, touch, hold hands, or give hugs. Cultural rites of passage were cancelled: no weddings, no graduations, no baby-naming ceremonies. Funerals were tiny or forbidden.

Health-care and personal-support workers were in constant danger. In the grueling atmosphere, charged with fear and death, they suffered tremendous exhaustion and depression, both physical and mental. Russia began an online Memory List (Список памяти) of caregivers who had died of their work. By early 2021, it featured more than one thousand names, with ages, professions, and locations. On 1 February 2021, a group in India, led by Kolkata doctors, started a virtual "National COVID Memorial," with photos and short tributes open to anyone, caregivers, and family members. The founders gathered stories of the poor and destitute, saying that no one should be left out and observing that the acute pain of the pandemic may alter traditions of mourning forever. In March 2021, on the first anniversary of a COVID-19 death, activists in Prague chalked more than 22,000 white crosses on the cobblestones of the Old Town Square, marking the number of Czech deaths and protesting the lax response that had placed its statistics among the world's worst. When on 24 May 2020, the United States reached the grim milestone of one hundred thousand lives lost, the *New York Times* attempted to gather all their names, ages, and occupations. By early 2022, the number of American dead had swollen more than eight-fold.

The passing of certain individuals personalized the crisis that might otherwise seem remote. The United Kingdom witnessed an outpouring of grief, on 2 February 2021, with the death of their darling Captain Sir Thomas Moore, the 101-year-old World War II veteran. He had captured the nation's heart – and a knighthood – by doing laps of his garden with his walker to raise funds for health-care workers. Canada's CBC News frequently devoted short episodes to celebrating lives of ordinary citizens, letting the bereaved express sadness or anger, and inviting a single tale to represent a multitude. The stories were archived at a CBC "Lives Remembered" web page. Later in the year, American media reported extensively on various COVID-19 deaths, including that of the seemingly indestructible eighty-three-year-old general and statesman Colin Powell.

In the Americas and Europe, scholars observed that depression, stress, and anxiety were on the rise, and substance abuse, of alcohol, cigarettes, and illicit drugs, was increasing too. Already an epidemic, opioid deaths steadily increased. Concerns over a possible rash of suicide appeared in spring 2020; by autumn, several reports had confirmed its reality with prevalence among the socioeconomically disadvantaged. Dentists observed damage from grinding teeth. Some observers had jokingly predicted a baby boom from the repeated lockdowns – as had followed lengthy power outages. But, in fact, the opposite was true, and psychologists described the dampening effect that isolation, anxiety, and financial insecurity had on sexuality, libido, and the desire to bring children into a gloomy world.

Domestic violence also increased globally. Confinement and financial deprivation provoked anger and frustration. Some people (most often men) were said to have suffered massive ego distortion when they could no longer provide for their families. Unable to leave home, they responded to the surging anxiety by abusing their spouses and children. Increased numbers of domestic assaults began appearing in early 2021. Calls to women's shelters in Canada doubled during early COVID-19. Femicide rose everywhere in what scholars have called a "parallel calamity."

When children contracted the virus, they were generally spared severe symptoms; however, all children were affected by the closure of schools and isolation from family and playmates. Although teachers struggled valiantly to deliver online education, it was an inadequate substitute for the in-person experience. Furthermore, the distress was uneven and followed social-class lines; some households, especially those with several children, did not have the necessary resources, time, or patience to help with the process. When schools opened with rules for masking, hygiene, and distancing, many were surprised to discover that even very small schoolchildren did their best to comply. Was it from fear? A February 2021 study from Toronto's Hospital for Sick Children found that COVID-19 placed a troublesome burden on child and youth mental health. Some experts insisted that negative impacts of the pandemic would last a lifetime.

The financial trauma of lockdowns on individuals was huge and has not yet been fully calculated. Repeat closures of nonessential businesses meant that owners could not pay employees, cover rent, or feed families; many failed. Homelessness increased. Like the shelters, food banks saw a massive rise in clientele, while provisions became scarce. Some restaurants began offering takeout options; others devised minimally priced meals for students or the homeless, helping to maintain morale of staffers who had to become volunteers but still wanted to work.

Travel restrictions and border closures gutted the aviation industry, and airline companies appealed for government bailouts to stay aloft. Countries that relied heavily on tourism suffered proportionately greater economic collapse. Not flying meant that grounded pilots had to repeatedly retrain virtually to retain their credentials. Would-be travellers wondered how the industry could assure future safety of aircraft and pilots, raising more doubts about how quickly tourism would resume. A kind of flight hesitancy arose, not only over safety, but because of worries that something pandemic-related might prevent a planned return home. Adding to the skepticism, the notorious Boeing 737 MAX, which had crashed twice since October 2018, was cleared to fly again. It seemed impossible that the imagined postpandemic world would see a rapid return to previous normal travel patterns.

LONG-TERM CARE

The first wave had demonstrated the extreme vulnerability of long-term-care facilities (LTCs), especially those for the aged and disabled, but the later waves brought yet more carnage. It seemed that neither advance warnings of successive waves nor time itself had prompted compensatory staffing or procedures to repair the poor conditions of architecture, ventilation, and crowding. Not all residents of LTCs were elderly, but they were always frail, often disabled with the chronic health issues that had forced them out of their own homes, making them exceptionally vulnerable to infection.

The report of the military intervention during the first wave of the pandemic was scathing (chapter 2). By September 2020, the Canadian Institute for Health Information (CIHI) began gathering information on ownership of the more than two thousand LTCs in Canada. Those in the Territories, Newfoundland and Labrador, Saskatchewan, and Quebec were mostly or entirely owned by the public. Elsewhere most LTCs were owned privately for profit, or in a mixed private-public arrangement; Ontario, Nova Scotia, and New Brunswick had the fewest LTCs in public ownership at 16, 14, and 0 per cent, respectively. The analysis showed that private, for-profit facilities fared worst in the pandemic.

Provincial governments also tracked outbreaks of COVID-19 in LTC homes, charting the number of cases and deaths. The results were staggering. Health-care staff from acute-care hospitals, already reduced in numbers by illness or quarantines, were transferred to help in LTC homes. For example, an outbreak beginning on 6 January 2021 in the privately owned, Roberta Place of Barrie, Ontario, became notorious when the virus reached 127 (91 per cent) of its 140 residents, killing sixty-nine, and infecting more than a hundred staff members with the more contagious UK (*alpha*) variant. Media focus on this dreadful situation and a threatened lawsuit tended to shelter the public from the stress and ongoing hardship in other care homes where the brutal effects were also keenly felt, and many died.

Canada's LTC statistics during COVID-19 continued to top OECD averages. Several provincial governments had abandoned inspections prior to the pandemic. The grim LTC figures were poised to join an end-of-pandemic analysis that would force a national confrontation with our habits of elder care and the various forms of LTC ownership, as described in an acclaimed book by health journalist, André Picard, released in early March 2021.

MANAGING THE SPREAD (OR NOT)

During outbreaks, regional health units could be overwhelmed. When numbers of cases exploded, the list of contacts ballooned exponentially. Under those conditions, some units abandoned contact tracing simply to monitor actual cases. To illustrate how the work quickly became unwieldy take the example of the relatively spared region of Kingston, Ontario, where I was volunteering as a contact tracer (chapter 6). One day near Christmas 2020 saw twenty-four new cases; adding those to the other active cases that had been accumulating resulted in the need for more than eight hundred phone calls to contacts in a single day.

This message was difficult for the public to grasp, especially for youth, tired of restrictions and hankering to socialize. Most people adhered to recommendations, though often they were depressed and lonely. Outrage greeted politicians who exempted themselves from the stay-at-home orders and quietly went off on winter vacations in sunny destinations. Several politicians and executives had to resign or accept demotions: seven in Alberta, at least four in Ontario. Their shame made international headlines, none quite as spectacular as those of US senator Ted Cruz or British Prime Minister Boris Johnson. Cruz took a one-day escape to Cancun during a winter storm that had left millions of fellow Texans without power or water. Johnson and dozens of colleagues partied at Downing Street during lockdowns; his Partygate lies sparked unheeded demands for his resignation. People suffering from disease or obediently accepting privation were disgusted by the flagrant hypocrisy, which did nothing to encourage their acceptance of ever greater sacrifices. Beyond the political sphere, in January 2022, global attention followed the case of a famous, unvaccinated tennis player who had believed himself exempt from Australia's strict rules; he was deported.

In later waves, the demographics of COVID-19 shifted slightly but maintained trends seen at the outset (figure 9.1). More cases were seen in the under forty age group – people who were not likely to suffer much with infection, if at all, and who were less likely to be vaccinated; meanwhile the greatest number of deaths continued to accumulate among the elderly. Already by early February 2021, Canadians over seventy years of age made up more than 90 per cent of deaths, a percentage that slowly drifted downward over the year to about 80 per cent, as the disease began to claim lives from growing numbers of younger patients. Gender distribution was equal in all age groups. This same pattern was observed around the world, although in countries where the elderly lived at home, proportionately fewer died.

As had been seen in the first wave, stringent lockdowns were most effective. The Oxford Stringency Index now has data from more than 180 countries. The team had refined its tool with more finely grained information of states or provinces within countries (chapter 6). Comparing the stringency index with numbers of cases and deaths simply confirms what was already known: the tighter and longer the lockdown, the better the outcome – especially where financial support and more active protection were available. The diverging observations held true even within countries. By early 2021, it became clear that regions that had been slow to react to the arrival of the virus suffered the most. The *Globe and Mail* published a report on 2 January 2021, using the stringency index, to show how delay and laxity correlated with far worse hospitalization rates. The same was observed in the Canadian West throughout that year (chapter 9 and below).

OTHER DISASTERS

The march of world events might have been eclipsed by the pandemic, but it did not cease. While this is a history of a disease, we cannot pass without mention of the many other catastrophes that occurred during the pandemic. They provided ideal conditions for the infection to spread, challenging already strapped medical facilities, especially in developing countries. Chief among them were the struggles of Lebanon after its devastating explosion on 4 August 2020, and those of Yemen crushed for more than a decade by the ongoing travails of war, famine, and cholera, where the case fatality rate of COVID-19 is as high as 30 per cent. In mid-August 2021, Haiti suffered another massive earthquake with more than two thousand dead, twelve thousand injured, and up to one hundred thousand buildings damaged or destroyed, including many healthcare facilities. The already limited COVID-19 testing and cases numbers in those places simply faded into the maw of general suffering; chances of survival diminished.

Several elections around the world generated massive crowds, gatherings that were frequently recognized as superspreader events, where regions were indeed capable of documenting the harm. In regions unable to perform adequate testing, the consequences were absorbed by the chaotic vortex of illness and deaths. Among them figured Donald Trump's crowded rallies both before and following the 3 November 2020 election (which he lost, although he continued to deny it). The indifferent reaction of his administration to the illegal activities of armed, white protesters, without masks, stood in sharp contrast to its intolerant, violent crackdowns on peaceful demonstrations for Black Lives Matter, during which most protestors wore masks. By March 2021, increased rates of infection in American counties that had hosted Trump rallies had been confirmed, and states that had voted Republican increasingly came to dominate the American pandemic (chapter 9).

Political upheaval in Belarus followed the contested 9 August 2020 election, considered fraudulent by massive numbers of citizens and many democratic countries. As its authoritarian leader claimed a sixth term, arrests, deaths, and exile of opposition leaders soon followed, while protesters, with and without masks, filled the streets. Police used the pandemic, the existence of which the president had denied, as a pretext to crack down violently. Well into 2021, both the protests and the vicious response continued, with chanting, rotating strikes, multiple arrests, and heavy fines for gatherings or wearing clothing deemed illegal.

In January 2021, massive protests also took place in Russia over the poisoning, arrest, and incarceration of opposition leader Alexei Navalny. As in Belarus, forces of order used the COVID-19 restrictions to crush demonstrators, jailing thousands. On 1 February 2021, the Myanmar military staged a coup against the ruling party that had just won a landslide victory, jailing its leaders, including Aung San Suu Kyi. There protests began slowly, led by doctors and nurses, gathering ever larger crowds; they were met with growing force, until three weeks later, police began firing on citizens and killing unarmed demonstrators. The death toll reached at least 150 by late March, and the protests continued. In Kazakhstan demonstrations arose in January 2022, triggered by rising fuel prices and government corruption. They resulted in the arrival of Russian troops, an estimated twelve thousand arrests, and 160 deaths; within a few days COVID-19 cases soared.

Natural disasters also complicated the ability of authorities to control the pandemic, and 2020 and 2021 seemed to witness an inordinate number, many related to climate change. Australian bushfires continued into March 2020. By late summer, wildfires were ravaging California. The 2021 wildfires in British Columbia destroyed almost one hundred thousand hectares of forest and incinerated the entire village of Lytton, BC, the day after it posted the national record temperature of 49.6C. Deadly floods took place in Afghanistan, Indonesia, France, Australia, and the lower mainland of British Columbia. Earthquakes and volcanoes ravaged Turkey, Greece, Spain (La Palma), and Tonga; the latter provoked a tsunami. Hurricanes, typhoons, or cyclones destroyed swaths of Bangladesh, the Philippines, Central America, the Caribbean, France, Louisiana, and Texas. Somalia and other African nations encountered the worst infestation of locusts in a quarter century and faced an emergency of food insecurity.

Conflict provides an ideal breeding ground for infection. As the American troops scrambled to pull out of Afghanistan by 31 August 2021, fulfilling Donald Trump's commitment to the Taliban, desperate, panicked people crowded the airport where they were attacked by terrorists. Their plight made recommendations for social-distancing measures against COVID-19 seem hopeless if not ridiculous. Afghan hunger is increasing; aid is limited, partly because of the ambiguity surrounding diplomatic relations with the Taliban. No one has confidence in the tiny COVID-19 numbers emerging from that impoverished nation; the pandemic may be the least of its worries. By the end of 2021, Russia was amassing troops and tanks along the border with Ukraine and had placed missiles in Belarus, prompting North Atlantic Treaty Organization (NATO) members to send arms and recall diplomats from Kiev. War was looming for weeks until it exploded in late February 2022.

Meanwhile, refugees fleeing conflict in Syria, Venezuela, and Yemen, as well as the Rohinga of Myanmar, were especially vulnerable to the pandemic because they lacked resources for basic hygiene and had no ability to track the disease. Conditions for peoples on the move are never safe, and with several developed nations, such as the United States, EU members, and the UK tightening borders, their lives are even more at risk. In late 2021, as a distraction and in revenge for foreign sanctions, Belarus manufactured a brutal refugee crisis along its border with Poland: cold, starving migrants were denied medical care; some were deported, and reports arose of deaths and mass murder. No one is confident that the pandemic numbers from that country bear any resemblance to reality. The wages of COVID-19 in refugee camps – an unnatural disaster – are yet to be determined.

Beyond the Numbers

HEALTH VERSUS WEALTH: A FALSE DICHOTOMY

At the outset, COVID-19 was expected to push more than one hundred million people into poverty. According to the International Monetary Fund (IMF), the negative growth of world GDP during 2020 was about minus 4.5 per cent, with wide variation, ranging from minus 7.2 per cent in the Euro zone, minus 10 per cent in the UK, minus 5.5 per cent in Canada, and minus 3.5 per cent in the United States. Within each country, youth, women, and uneducated workers were hit hardest. The economic loss was not as great in developing nations, but the effects are predicted to last much longer. Less wealthy countries could barely generate enough resources to sustain themselves during the crisis and relied on IMF and World Bank assistance.

As of 30 June 2020, the World Bank had increased its commitments from an average of about US$63.3 billion per year over the preceding four years with another US$77 billion; it planned to deploy up to US$160 billion by June 2021. The IMF also responded. Between March 2020 and March 2021, the IMF had made US$118 billion available to eighty-seven countries in Asia, Africa, and Latin America. The amount was considerably more than the US$70 billion it had made available to eight countries in the 2018–19 year. It also cancelled or extended debt repayments for twenty-eight of the world's poorest nations. In August 2021, it allocated US$650 billion in special drawing rights, the largest amount ever, to help pay for health care and buy vaccines. Nevertheless, some claim that it should do more.

Opposing the concept of financial support was a conservative view that pitted economic stability against health. Adherents pointed to rising deficits and the exorbitant financial cost of public-health measures, such as closing nonessential businesses, which would lead to losses of income, jobs, and tax revenue. They resisted pouring more public funds into supporting the sick, the quarantined, and the poor. "We can't afford it!" they argued, pointing to the dilemma of people having to choose between paying rent or buying food: "Let them work!" Sometimes they characterized restrictions as an infringement of personal freedom. They also pointed to the dramatic drop in GDP at home and around the world. Political leaders were often swayed by these concerns coming from a noisy sector of usually well-off voters.

But the experience in East Asia and the Pacific proclaimed this debate between health and wealth as a false dichotomy. A mix of democratic and authoritarian countries in the region, such as Taiwan, New Zealand, China, and Vietnam (at least initially), had controlled the pandemic with rigorous measures, masking, contact tracing, high public compliance, together with political support. And they had maintained their economies. Pandemic control preserved and enhanced economic activity.

Many economists in the West agreed. They rejected the political argument that relief measures were unaffordable and described objections as a political device playing to temporary circumstances. Take the example of Switzerland and its highly fragmented system of government where tepid public-health measures had allowed COVID-19 to spiral to one of the highest rates in the world. In early November 2020, eighty Swiss-based professors of economics wrote an open letter to the federal government demanding that it impose a strict lockdown. Admitting that expenses would be incurred in the short term, the professors described the much greater costs of a lengthy, out-of-control pandemic, with the additional grievous expense of many more lives lost. "Saving the economy requires controlling the pandemic," they wrote, ending with an extensive international bibliography. Among their references figured the March 2020 anthology, edited by Richard Baldwin and Beatrice Weder de Mauro: *Mitigating the COVID-19 Economic Crisis*. Their subtitle says it all: *Act Fast and Do Whatever It Takes*.

Writing in the *New York Times* on 7 February 2021, the Nobel laureate in economics, Paul Krugman, admitted that the pandemic had had a huge effect on the world economy, but it was not a banking crisis and not a "conventional recession"; it was a natural disaster, like an earthquake, a tsunami, an asteroid crash, or a war – on a global scale. It was wrong, he wrote, to refer to these massive national and international expenditures as "stimulus packages." Wherever they were made, in either advanced or developing economies, they were about basic survival and relief, helping people get to the other side.

As the pandemic wore on, the economic crunch was felt differentially between groups within countries. White (and white-collar) workers who could work from home did well economically and could afford to pay the soaring prices for new houses where they could live and work more comfortably. At the other extreme, the pandemic was devastating for minimum wage workers, racialized communities, women, and single-parent families. It also impacted high-contact businesses, like restaurants and retail; politicians felt the heat from those owners and customers, making them hesitate to follow public-health advice for lockdowns. Some analysts recognize the pandemic as a crisis that worsened polarization, and it made wide sectors of the population susceptible to the specious theories of the infodemic.

Despite all the evidence to the contrary, and unlike economists, politicians sometimes view health versus wealth as a true choice; and it is politicians who control public resources and laws. The attitude makes them wary of lockdowns and of doling out money to alleviate the stress. On one hand, strict lockdowns bring immediate hardship, financially and psychologically; they are unpopular with some voters, often those having the resources to complain loudly and donate to campaigns. On the other hand, some political leaders are conservative by name and nature: they genuinely view saving money as a form of fiscal responsibility. Therefore, they are duty bound to limit public spending on those who have been disadvantaged by control measures, either through lost business or employment, or both, or because they cannot afford mandated isolation or quarantine. Benefit payments, sick leave, and quarantine hotels could ease those burdens, but they would cost even more. People might clamor for support but providing funds and then dealing with the resultant debt proves controversial. Those on the right threaten unpopular tax hikes if support should be given. Those on the left, or in the middle, argue that support funds are essential and can be taxable; consequently, debt would decline as a share of GDP and taxes would not have to rise. Politicians also occasionally argue that the donated funds might keep afloat some poorly run businesses that deserve to die.

THE PACKAGES

While waiting for recovery, most countries extended deadlines or beefed up their national relief packages to sustain businesses and relieve furloughed workers: for example, Germany (late August 2020); France (October 2020); UK (January 2021).

The United States, however, was frozen. The formidable CARES Act package of late March 2020 had provided funds to households with unemployed members, but the distribution was uneven, and the money was running out by the summer with little change in high unemployment or need. Owing to political fighting, bizarre arrangements resulted, such as giving $1,200 to every adult and $500 to every child whatever their income status (chapter 6). Before and after the election, no new package would appear until late December 2020. It was too modest.

Following his inauguration in January 2021, Joe Biden's new administration introduced the *Rescue Plan Act*, a new package of US$1.9 trillion, an amount so large that it found two congressional opponents even within his own Democratic party. Targeting lower income people, it promised to help raise living standards and reduce poverty. His political opponents argued that it was too expensive and too partisan, most need being in states that had voted for Biden. However, it pleased the general public, narrowly passed the Senate by one vote (with no Republican support), and was signed on 11 March 2021. At the time of writing, its effects are still being analyzed; a late 2021 report suggested that it had fueled inflation and price hikes. That conclusion was in dispute, however, because as the economy recovered, inflation was happening everywhere with or without benefit supports (chapter 12).

IN CANADA

In Canada, too, public actions were simultaneously protested for being both too little and too large. By the end of 2020, Canada's supports had risen to about 10 per cent of GDP, and new envelopes were added with increased amounts, reduced taxes, and extended benefits, especially in advance of what promised to be a severe second wave. It forgave repayment of some benefits that had been given in error to the great relief of recipients who had already spent the money on rent, food, heat, and light. The supports were coming to an end by late 2021 when the *omicron* variant began its relentless assault: a new benefit package was established (chapter 12). Opposition leaders on the right claimed that too much money had been spent with too little effect; those on the left pushed for more relief.

Political differences played a huge role in the spread of the disease. Typifying the debate were conflicts pitting provincial leaders against public-health experts, federal colleagues, opposition counterparts, and large groups of doctors. During the autumn 2020 surge, Alberta Premier Jason Kenney wrongly contended that "sweeping lockdowns" elsewhere were "indiscriminately violating people's rights and destroying livelihoods." Outraged, more than three hundred physicians signed their third letter to him describing their exhaustion, the potential collapse of services, and the need for a lockdown. On 24 November 2020, the premier finally announced a second state of emergency (not a lockdown, mind you). In early December, he began issuing tepid recommendations with curfews and orders against large gatherings in bars, restaurants, and churches. Until 8 December 2020, Alberta had been the only province without a mask mandate.

Nevertheless, numerous Albertans objected to these weak measures, staging demonstrations over the perceived violation of their freedom,

Beyond the Numbers

and church leaders insisted on holding in-person services. Some were charged with contravening the *Public Health Act*. Observers had long detected a fraught relationship between Alberta's chief medical officer, Dr Deena Hinshaw, and the government. The media aired twenty leaked audiotapes of confidential meetings implying that the politicians were ignoring and micromanaging the public-health experts; it provoked embarrassment and a sense of betrayal for all concerned. But to the dismay of physicians, Alberta eased restrictions over the 2021 New Year, lifting them completely by early February as the caseload was declining; however, the so-called decline in new cases had dropped only to just below where it had been at the declaration of the state of emergency, and more contagious variants were beginning to circulate. In late January 2021, biologist Gosia Gasperowicz calculated that a stiff seven-week lockdown could eradicate the virus in Alberta, as it had done in New Zealand without a vaccine. Her advice was ignored.

As much as people hate the sanitary protocols, the measures make a huge difference to disease control. Alberta's Premier Kenney was viewed as accountable for the disastrous open-for-summer decision (chapter 9); however, as case numbers soared, he spent many weeks in hiding, until he finally issued an apology in mid-September 2021, admitting that the decisions were not only premature, but "wrong."

Similarly, Saskatchewan Premier Scott Moe claimed that lockdowns were merely a "stopgap" measure, accusing those calling for them to be the privileged, wealthier few "who could work from home." Only a vaccine, he said, would stop the pandemic, and businesses needed to stay open while waiting. On 9 February 2021, the New Democratic Party (NDP) opposition leader, physician Ryan Meilli, who had been calling for action since November, presented a petition urging lockdown, signed by four hundred health-care professionals working on the front lines and definitely not from home. Unlike Kenney, Moe refused to apologize. Later, Meili criticized Moe for ongoing laxity in refusing to limit Thanksgiving 2021 gatherings and in hesitating to invite federal help; he demanded more transparency and accused the premier of placing "politics before people." Like Meilli, opposition leaders in all three prairie provinces were strongly critical of their governments' actions.

Ontario also saw a similar battle between public-health recommendations and political expedience. As the second-wave caseload began to turn a corner in early February 2021, Premier Doug Ford announced a gradual loosening of the province's admittedly tighter restrictions, just as the variants began circulating. The result was a devastating third wave.

155

By April 2021, his desperate government reversed course and tightened restrictions. He even closed playgrounds and issued police with powers to randomly stop and question people disobeying curfews. The police, however, refused to cooperate saying that the measures were against the *Charter of Rights and Freedom* and that they had too much other work to do. On 22 April 2021, Ford walked back some measures and apologized: "Simply put," he said, "we got it wrong." Nevertheless, the province maintained its masking and social-distancing rules during the vaccination campaign; its fourth wave was less severe. Consequently, observers were alarmed when his announcement to ease restrictions in late October 2021 came with a foolish prediction that vaccine mandates and proofs would no longer be necessary in less than three months. It was a sop to anti-vaxxers, heralding a decline in vaccinations – why bother if the shot soon won't be necessary? – and a replay of Kenney's open-for-summer fiasco.

No provisions had been made for sick pay or quarantine pay in any of these three provinces where cases and contacts were ordered to stay home for up to two weeks. The leaders seemed to fear a voter backlash from angry business owners and the unemployed more than they feared the virus. An economist-father and physician-son team, Munir and Hasan Sheikh, writing in the *Globe* (6 April 2021) used prospect theory to analyze the politicians' failure to comprehend human behaviour and their miscalculation over where most votes would lie. The result of the political dithering was delay and ineffectual measures, which came too late and were lifted too soon, bringing a new surge, sometimes with variants.

While the scientific advice within Canada had been the same across the country, the policy differences between regions became flagrant and are expressed in variable numbers of the sick and the dead (figure 9.4).

CANADIAN ELECTIONS

COVID-19 participated in elections everywhere, causing delays, inviting protests, and leading to accusations of added risk to voters. According to the International Institute for Democracy and Electoral Assistance (IDEA), most postponements took place in early 2020; however, the anticipated rise in infection usually did not occur. Voter turnout seemed to decline in most places, although the United States saw an increase.

By mid-2022, Canada had seen six provincial elections during the pandemic: New Brunswick (14 September 2020); Saskatchewan (26 October);

British Columbia (8 November); Newfoundland (25 March 2021); Nova Scotia (17 August 2021); Ontario (2 June 2022). Despite predictions that the elections would spread the disease, none appeared to do so. New Brunswick delayed its vote. Only Newfoundland with its mail-in solution was seriously disrupted. Incumbents were re-elected in all but Nova Scotia – a surprise as its pandemic had been well managed. Leaders and voters maintained a posture of respect for the pandemic, but the quality of COVID-19 management became a major plank in all campaigns. The criticisms meant that the federal minority government teetered along under the ever-present threat of an election.

But the federal government did not fall; it chose, instead, to fold. In mid-August 2021, Prime Minister Justin Trudeau called an election for 20 September. Until early August, public opinion of his Liberal government had been riding high, owing to general satisfaction with pandemic management. To justify his decision, he claimed that the last election had been held before the existence of COVID-19 and that a renewed mandate should precede the important decisions that lay ahead. Indeed, there had been no shortage of criticisms, and Liberal support began to trend down. It was obvious that his party hoped for and even expected a majority. All opposition parties complained that the election call was wasteful and tried to make it a greater issue than pandemic management. "He just wants a majority," sneered one Conservative ad, as if that party would never want a majority for itself. Then the Afghanistan crisis emerged, and the government was confronted with the emergency evacuation of Canadians and Afghan citizens who had worked for Canada prior to its own pullout under conservative Prime Minister Harper in 2014. Having no troops on the ground, Canada struggled to evacuate a few thousand people, using charter planes and help from allies, and it committed to resettling tens of thousands more when they could leave.

Polls showed Liberal support eroding over the month of campaigning, but the party used conservative ambivalence over vaccines as a wedge issue. To the exasperation of public-health experts, the People's Party of Canada (PPC), led by libertarian Maxime Bernier held many crowded rallies, attracting media attention. Deliberately flaunting sanitary measures, Bernier refused vaccination and masks, and he reveled in being charged with violating laws. Some argued that in grabbing 5 per cent of ballots, the extreme PPC views helped the federal Liberals by splitting the right-wing vote.

The act of voting itself was complicated because of pandemic restrictions. Expenses were greater than usual. Extra staff was needed. My husband and I volunteered to help work a site with four polling stations in our rural Ontario riding. Most people were polite and patiently stood, socially distanced, in the line that I was managing outside. Only a handful angrily refused to put on a mask, even for the few minutes that voting would take. Nevertheless, with or without a mask, they had the right to vote. When an unmasked person arrived, everyone else had to wait outside while the hall cleared of other voters, the refuseniks cast their ballots, departed, and the whole station was disinfected. The line grew long during the wait. To my amazement, one rude objector claimed to be a nurse. Little doubt where she sent her ballot. In our location, 10 per cent of voters chose PPC, more than double the national average.

In the end, another Liberal minority government was returned with a seat distribution almost identical to what it had been in 2019. The election was fought and won on pandemic performance, but the result entrenched the idea that it had been a frivolous waste of time and money.

THE INFODEMIC AND OTHER DISTORTIONS

The word "infodemic" (from "information" and "epidemic") is said to have been coined in 2003 during SARS; however, it is associated with COVID-19 ever since WHO Director General Tedros began using it in his speeches as early as February 2020. Others prefer to call it a "disinfodemic" since usually it signifies widespread information that is incorrect. One important vector is social media. It is part of the "syndemic" of simultaneous problems that defy biomedical models and responses. The WHO posted a statement signed by 132 countries warning of the infodemic, and it has been hosting conferences on its management since June 2020. It recognizes an irony that the very technologies and social media that help keep people safe, informed, and connected, are the same tools that enable and amplify the infodemic, preying on fear, undermining responses, and jeopardizing measures to control the pandemic (figure 10.1).

False rumours are nothing new in pandemics. Historians of 1918 influenza have shown how articles, posing as scientific reports, were planted in newspapers to promote fake cures. What is new during COVID-19 is the ubiquity of social media, the echo chambers that they create, and postmodern skepticism. People believe that they can easily do their own research. They inadvertently select voices and ideas that match their hopes and expectations, in blithe disregard for credentials, or

Figure 10.1 | Editorial cartoon by Bill Bramhall, *New York Daily News*, 21 October 2021

even with a loathing of expertise. Cartoonists gleefully sketched graves bearing epitaphs: "I did my own research." The concept of "epistemic trespassing" – spewing of so-called knowledge – applies to those who disseminate false information. For doctors and nurses treating the sick, and for public-health workers trying to stop the spread, the infodemic and the epistemic trespassers were actively working against them.

On 25 February 2021, the Washington-based Atlantic Council published *Weaponized: How Rumors about COVID-19's Origins Led to a Narrative Arms Race*. The nonpartisan think tank, founded in 1961, aims to shape solutions to global challenges. Its digital research lab, operating since 2016, exposed the extent of disinformation, conspiracy theories, and lies that had arisen about the pandemic from as early as 5 January 2020. Perceived enemies charged each other with having engineered the virus. Sources in Hong Kong, the United States, and the United Kingdom accused China. But China, Russia, and Iran accused the United States. By June 2020, the European Commission had accused both Beijing and Moscow of trying to undermine western democracies in claiming that SARS-COV-2 had been made in an American lab. Republicans within the United States accused

Democrat-directed funds from within their own nation. According to the report, superspreaders of lies included Russian Igor Nikulin, who claimed to be a bioweapons expert, and the American lawyer-and-conspiracy-theorist, Francis Boyle. They formed a pair of almost equal and opposite disseminators, each blaming the other's country.

The inflammatory lies of the infodemic may have emerged in individual Twitter or Facebook accounts, but they were picked up by political entities and media outlets, and then fed as news to unsuspecting readers of state-sponsored journalism where they found their way into normally respectable outlets in places such as Finland, New Zealand, and Europe.

Canadians were shocked to find their country named in this infodemic report. One theory had appeared on 26 January 2020 at Great Game India, an online source, devoted to conspiracy theories and laced with errors. It contended that a Chinese-born scientist, named in the report, who was working at Winnipeg's National Microbiology Laboratory, had infiltrated the lab with spies and stolen an original SARS coronavirus for Wuhan where it would be engineered, through gain-of-function work, as a bioweapon. The report cited Francis Boyle. In fact, the scientist involved, who had helped create a treatment for Ebola, had indeed broken policy rules when she sent samples of different pathogenic viruses to the Wuhan virology lab in early 2019. She and her biologist husband had been dismissed in July 2019 well before the pandemic, and an RCMP investigation was launched. Both the Public Health Agency and the RCMP claim that the dismissals had nothing to do with coronavirus, although activists and opposition politicians continue to demand more details. The grain of plausibility keeps these theories alive.

Canadians also learned that a Montreal-based website (globalresearch.ca), purporting to be the arm of a think tank, had been an early participant in the infodemic. Founded in 2001 by Michel Chossudovsky, a former University of Ottawa economics professor and conspiracy theorist, the site hosted two reports by in-house writer Lawrence Devlin "Larry" Romanoff, on 17 February and 4 March 2020, both citing Igor Nikulin and accusing the United States of having engineered the virus and its military of disseminating it. The charge was picked up and retweeted by Zhao Lijian, a spokesperson for the Chinese Foreign Ministry. The same website was outed as having promoted Kremlin-sponsored interference in the 2016 American election; in 2017, it was reported to have aired Russian-sponsored disinformation in support of Syria's Assad regime.

Not all the disinformation was to do with theories of weaponized viruses. Some came from normally respected nation states trying to save face over their pandemic management. Richard Horton, editor of the *Lancet*, was horrified to find his own words distorted and taken out of context by the UK government trying to justify its early mistakes.

When in late 2020 countries began requiring a negative COVID-19 test for entry or travel, fake test kits and results became available for sale. FDA and other entities issued warnings against them, adding them to its warnings against fraudulent remedies. By early February 2021, border-control technology had been upgraded to detect fakes, and travellers to Canada had been fined up to $7,000 for attempting deception. Soon fake vaccine certificates would become available.

ECONOMIC RECOVERY?

Meanwhile, 2021 was predicted to see 6 per cent growth, and the economy in many nations began to revive. China was already on the way to recovery by the end of 2020 – and it did not escape notice that its restrictions had been among the tightest and the longest, while its financial supports and expenditures had been considerable (chapter 6).

In 2021, the United States and Canada both saw a spectacular year in recovery of unemployment and growth. Krugman credited the government support. Nevertheless, inflation was high, and some shortages and supply-chain problems persisted. For example, the auto industry suffered a slowdown owing to a global shortage of microchips; this situation boosted the market for used cars. The pattern of recovery displayed two very different shapes between countries and within single countries: one, called v-shaped, with a sharp decline and an equally sharp recovery that was seen in GDP, stocks, and trade. The other, called K-shaped, concerning jobs and income, featured two lines: one trending up (for wealthy economies and white-collar workers) and another trending down (for poor nations, blue-collar workers, service industries, and women). Recovery for the latter group would take a longer time.

Even as the economy roars back, we cannot know how long the scarring from personal tragedy or financial disaster will last. It remains for the future analysts of COVID-19 to show us if politicians facing elections were quicker or slower to disburse funds, and what difference, if any, resulted from where they stood on the political spectrum. Sadly, those actions, which had valiantly controlled pandemics of the past, were viewed as little more than Brownian movement. What the world wanted was a fix. The world wanted vaccine.

11

Vaccine Wars

It is a marvel of modern biotechnology that we have any vaccines at all and that several versions appeared in less than a year from the first hints of the coming plague. But the rapid creation and testing of these effective vaccines turned out to have been the easy part (chapter 8). The rollout was far more painful, and it degenerated into useless, selfish squabbling within and across borders. Mass vaccination programs were cumbersome and patchy, even as they continued to provide vital research information. Public-health experts knew that some people would crave vaccine and snap it up if only they could get their hands on it. They also knew that others would hesitate and need convincing. Vaccine reluctance is a longstanding matter, but what came to be called vaccine nationalism was new.

VACCINE HESITANCY AND ITS ORIGINS

From the beginning, pandemic experts warned that even if vaccines could be created, they would have to confront traditions of hesitancy, entrenched more deeply in some places than others ever since vaccines first appeared more than two centuries ago. Experts dismiss the power of hesitancy at their peril.

Mothers always fret over pricking the tender skin of their pristine infants to introduce some mysterious, foreign substance that might bring harm. And some vaccines have indeed brought harm. Routine smallpox vaccination was eventually abandoned when it caused more adverse reactions than the number of infections it was preventing. Polio vaccine's famous Cutter incident of 1955 caused an outbreak of that very disease, affecting forty thousand American children, killing ten, and leaving two hundred with permanent damage. The 1976 swine flu campaign of President Gerald Ford vaccinated 45 million, but eventually, proved to have been unnecessary, and it left more than 350 citizens with Guillain-Barré paralysis. A 2009 vaccine campaign for vaccination against H1N1 influenza produced an outbreak of narcolepsy in Europe. These tragic outcomes became public-relations disasters.

Other earlier events contribute to vaccine hesitancy. Some are all too real, like the examples cited above; others are fraudulent. Nevertheless, they are amplified through social media. Among the fraudulent, is the influence of Andrew Wakefield whose now discredited 1998 article in *Lancet* (since retracted) wrongly implied that measles-mumps-rubella vaccine caused autism. Wakefield was eventually struck off the medical register in his native Britain, but he continues to campaign against vaccines, courting groups for civil liberties and alternative medicines. Thanks to him, the nonexistent link between vaccine and autism has become frustratingly persistent.

Other causes of vaccine hesitancy include the human-rights abuses incurred by medical experiments (not all involving vaccines) that were conducted on racialized, colonized, disabled, and other vulnerable people: the Tuskegee syphilis experiment left African Americans without penicillin for decades. Canadian trials cruelly starved Indigenous children with the deluded goal of understanding their physiology. It is no accident that general mistrust of authority and scientific expertise, including public health, remains far greater in communities grievously harmed in the past.

Similarly, mandatory legislation for vaccine often backfires. Democratic societies are allergic to the state telling its citizens what they can and cannot do with their own bodies and beliefs. Nevertheless, various countries have pursued such regulation, especially when challenged by diseases that can be prevented. In 1806, Napoleonic administrators ordered Italian states to vaccinate their citizens against smallpox and then demanded the same for students at home. The UK 1853 *Vaccination Act* was to apply to all babies at three months on pain of exorbitant fines or even imprisonment. Riots ensued. Canada too had seen vaccine riots long before COVID-19. For example, in Montreal during the 1885 smallpox epidemic, the largely anglophone, Protestant authorities tried to enforce vaccination on the mostly francophone, Catholic community.

In 2017, following a dreadful outbreak of measles, Italy made childhood vaccines mandatory. Once again, protestors took to the streets; many disobeyed the laws and refused. Rather than directly mandating vaccines, countries have found indirect carrot methods more effective. For example, in some Canadian provinces, access to public education or child-benefit payments is contingent upon proof of vaccination. Health-care workers and learners must receive vaccines for employment or study. Exceptions can be granted on compassionate grounds or for medical reasons and conscientious objection. Encouraging methods vary, even within countries, but they are rarely enforced until an outbreak draws attention to their existence.

In advance of the COVID-19 vaccines, countries had been gauging the degree of hesitancy within their populations. With the pandemic, it became a trope of media coverage that national and global safety would be achieved only with herd immunity at 60 or 70 per cent, or more, of the population. Anthony Fauci admitted in late December 2020 that he had been slowly moving the herd-immunity goalposts upwards to 85 per cent because he thought that the country was growing ready to hear it. Unless many were to suffer and die, at least that proportion of people everywhere must accept vaccination when it became possible and available.

A study published in the *Lancet* in September 2020, based on data gathered in 2015 and 2018, compared changing attitudes across 149 countries to vaccines. In general, there was an overall trend to greater acceptance. Countries in South America and Africa were most tolerant; North America and Oceania were moderately approving. Greater skepticism was found in Russia and its former satellites; however, France – the land of Louis Pasteur, inventor of rabies vaccine – was a stunning outlier of negativity within a generally accepting European Union.

France's vaccine hesitancy was known well before COVID-19 and had been studied medically and sociologically. Accentuated by conspiracy theories swirling on social media, it had reinvigorated roots in several events of the 1990s. For example, a surge in multiple sclerosis came hard upon a French campaign to introduce hepatitis B vaccine (although no causal link was established). Similarly, its tainted-blood scandal had resulted in HIV-infected product being used in transfusions, causing disease and death, and eventually sending doctors to jail. Some French citizens believed that vaccine, containing biological material, might induce infection; others thought it would introduce surveillance microchips or alter their DNA, turning people into monstrous, genetically modified organisms (GMOs). These fantasies were multiplied on social media and spread around the world through the infodemic.

Scholars also correlated hesitancy with extreme political views – both left and right – and with mistrust of government. Moscow journalist

Dasha Ryzhkova stated that Russian hesitancy stemmed from the long legacy of government lies that bred general skepticism. Citizens knew that Sputnik v had been approved by the president, but not by the public-health authorities or by the WHO. Putin himself waited seven months, until 23 March 2021, to take the shot himself, behind closed doors; the Kremlin declined to say which vaccine he received. On the other hand, informed observers complain that Russian and Chinese vaccines are reliable and subjected to even more stringent rules than others. The politically charged approval process, they contend, contributes to widespread hesitancy and frustrates the international rollout.

Everywhere – given tragedies of the past – groups identified as vulnerable – elderly, disabled, Indigenous, or racialized, those sequestered in long-term care, and even some health-care workers – imagined that they would be given top priority for vaccination, not as a gesture of support and sympathy, but as an experiment to assess safety before immunizing the population at large. They feared it was a conspiracy, a plot against second-class citizens.

VACCINE NATIONALISM

Following China and Russia's decisions to begin vaccinating citizens in summer 2020 before trials had been completed (chapter 8), the earliest vaccine released in the West was the product of Pfizer-BioNTech. Israel leapt ahead on a massive scale proportionately greater than any other country. Having suffered a severe second wave, it launched a concerted effort on 19 December 2020 to immunize its entire population. In the first five weeks after approval, 48 per cent of Israel's nine million eligible citizens had received a shot. How did it manage that feat? Israel had made a deal with Pfizer-BioNTech. For accelerated access, it would provide the company with anonymized, demographic data of its vaccinated people: their physical reactions and its effectiveness – age, sex, side effects, cases and deaths of COVID-19. This information would be of great value to the company as an extended phase-three trial. With a universal health-care system and resources for vaccination, it was easy to administer the vaccine and gather the data. Israel is said to have paid an inflated price for the privilege of obtaining so much vaccine early on. Its leaders reasoned that continued lockdowns and repeated surges would cost even more. Nevertheless, Israel was criticized for ignoring privacy issues, since it did not always obtain consent for passing on the anonymous data. Moreover, it was accused of vaccinating Jewish settlers, while ignoring others who dwelt in occupied Palestinian territories. In response, Israel claimed that those "others" were under a different health authority. Israel also refused to vaccinate health-care workers in the occupied territories and did not begin vaccinating Palestinian workers in its own territory until March 2021. The Palestinian authority could not afford the Pfizer vaccines. Consequently, on 11 January 2021, it granted an emergency licence to Russia's Sputnik V vaccine but feared shipments would be blocked by Israel. The first shipment of COVAX supplies, from the UN's sharing pool, arrived on 17 March 2021.

Countries with facilities for making vaccine wanted to ensure early access for their own citizens, although companies guaranteed that contracts for reserved doses elsewhere would be honoured. In May 2020, French President Emmanuel Macron summoned the top officials of Sanofi to complain that first doses of its as yet nonexistent vaccine would go to the United States because it had made the largest investment. Canadians bristled when their prime minister told them that they too would have to wait their turn. The world was having a rude awakening to the convolutions and interconnectedness of global pharmaceutical chains.

But the manufacturing companies, with their reassuring promises of millions of doses, seemed to have bitten off more than they could chew. Vaccines were not flooding into their client nations as expected. Political cartoonists had fun portraying Operation Warp Speed as a turtle.

Among the abandoned and delayed products are French vaccines from the Pasteur Institute and Sanofi. The former had tried unsuccessfully to adapt a measles vaccine and gave up. The latter modified a flu vaccine but announced it would not be available until the end of 2021, pending clinical trials, which, it claimed, were positive though incomplete. The news reports of "flops," "failures," and "humiliations" were viewed symbolically by French politicians as signs of national decline. The political leaders leaned on Sanofi in an unusual and possibly unprecedented manoeuvre, forcing the pharmaceutical company to agree to adapt its facilities in Frankfurt to making the rival Pfizer-BioNTech product – for French citizens. The urgent business of making vaccines had become a kind of Olympic blood sport.

By late 2020, the United States had suffered more COVID-19 cases and deaths than any country in the world. It began vaccinating with Pfizer-BioNTech vaccine on 14 December, followed soon by Moderna, both companies having several American manufacturing plants. The doses were purchased federally and sent out to the states to be given by hospitals, at least at first. When the Biden administration took over on 20 January 2021, it found "a mess": every state was different; at least twenty million doses were unaccounted for, probably owing to delays reporting injections; federal reserves of vaccine were low; and the flow of vaccine production was dwindling to a trickle. Nevertheless, by early February 2021, six weeks into vaccination, roughly 9 per cent of the American population had received at least one shot: 5.5 million people were fully vaccinated with two shots; another twenty million had had their first dose. Children and adolescents were not eligible because vaccines had not yet been tested in youth. By late May, 50 per cent of the population had received at least one dose, but then, the pace slowed and by early November, only 58 per cent had been fully vaccinated.

The United Kingdom, labouring under one of Europe's most severe second waves, was quick to vaccinate many senior citizens and health-care providers. It used both the Pfizer-BioNTech vaccine and the British Oxford-AstraZeneca vaccine, which it had been first to approve in late December 2020. Through its National Health Service (NHS), it targeted all people over age seventy and health-care workers, aiming for and achieving a vaccination rate of 380,000 jabs a day. That rate would bring complete coverage to fifteen million seniors and health-care workers by mid-February. Provided supplies continued, the NHS planned to cover the entire population in cohorts of decreasing age by autumn 2021; in November, a respectable 70 per cent were fully vaccinated. Vaccine loyalty in the UK contrasted with skepticism elsewhere. The media featured scenes of obedient, bemasked octogenarians quietly lining up for their jabs under the gothic arches of great cathedrals to the soft organ music of Bach, Handel, and Pachelbel. As its death toll topped 100,000, vaccine delivery was said to be the one thing that the UK got right.

To stretch scant resources and enhance protection, experts toyed with the idea of extending the period between the first and second shot – from the recommended three or four weeks (depending on the vaccine) to six or eight weeks, even six months. That way twice the number of people could be given at least partial immunity, and they could count on receiving second shots in the future when supply increased. At first, the pharma companies cautioned against this move, because their trials were based on the shorter interval between doses. No evidence supported whether (or not) a longer delay would be equally effective. Perhaps the companies also knew that future shipments were in jeopardy. Eventually, more reassuring data emerged to suggest that even a first dose brought significant protection for several months and that long delays between doses might be more beneficial, although short delays for the elderly were preferred.

Leaving aside the complexities of storage, transport, and freezers, countries also seemed baffled by who should administer vaccines and how to do it quickly and efficiently. France witnessed a turf war over allowing pharmacists to administer vaccine. Others argued that, rather than making people come to *vaccinodromes*, taking the vaccines into LTC facilities made more sense. Retired doctors in Canada, wanting to help ease strain on harried health-care workers, volunteered to give shots but were told to renew their medical licences and malpractice insurance at enormous cost. Communities began patchy conversion of unused gymnasia, hockey arenas, and factories – but the vaccines did not arrive.

The logistics of mass vaccination were so far in the past, so neglected, and so much out of mind that many privileged nations were utterly unprepared. In fact, they were less prepared than African countries that had recently managed to clamp down on Ebola fever several times. In mid-December 2020, the *New York Times* published a mind-boggling, historical essay that explained how, in 1947, the city had vaccinated six million against smallpox in under a month. Of course, back in 1947, smallpox vaccine was far from novel – and domestic supply was ample. But wow! How had they done it?

In mid-January 2021, a key Pfizer-BioNTech plant in Belgium had to shut down for structural changes that would eventually allow for an exponential increase in production. That factory was the source of vaccines destined for Canada. The numbers of imported doses fell dramatically and then stopped for all clients, except those in the United States. American factories were still churning out product, although not at the promised speed. All the while, Pfizer-BioNTech kept issuing reassurances that its commitments would be met by the promised dates, later in the spring.

One by one the makers of the shiny new vaccines missed deadlines and failed to meet targets: Pfizer-BioNTech, Moderna, Oxford-AstraZeneca, and even Sputnik V. Had they been unrealistic in their original planning? Were the scaled-up factories unequal to the demand? Was that a surprise? Or, had they known all along that hooking clients was more important than keeping promises? After all, they had created two-dose vaccines: if the first shot had already been given, the second was a guaranteed sale. Experts were asked if the second dose could reasonably come from a different vaccine. So much was unknown. Later still, in October 2021, experience demonstrated that different types of vaccines could be mixed with equal or even enhanced efficacy.

Lost in all the noise of vaccine envy and vaccine nationalism was the fact that all countries were suffering the same drop in deliveries. Italy was so angry over the issue that it threatened to sue Pfizer. Scientists behind the Oxford-AstraZeneca vaccine accused French President Macron of deliberately and unfairly discrediting its product to dampen market craving. The president of the European Union, Ursula von der Leyen, herself a physician, accused AstraZeneca of favouring the United Kingdom and neglecting its commitments. Meanwhile, developing nations were receiving no vaccines at all – because they could not afford them, and they (just like Canada) did not make them.

European outrage over vaccine shipments to the recently Brexit-ed UK prompted an extraordinary and unprecedented order on 30 January 2021: the EU would suspend all vaccine exports from factories on EU soil. Known euphemistically as the "transparency and authorisation mechanism," it was promptly criticized for being a shortsighted, counterproductive measure. Ending the pandemic depended on global distribution. The move betrayed the EU's commitments through the World Trade Organization to avoid such restrictions on matters pertaining to health. It was the flip side of the issues over PPE seen in spring 2020; this time, however, the makers were in Europe and America, not in China.

In dismay, Dr Tedros of the WHO lamented the vicious selfishness of vaccine nationalism. In a speech, delivered 18 January 2021, he said, "Even as they speak the language of equitable access, some countries and companies continue to prioritize bilateral deals, going around COVAX, driving up prices, and attempting to jump to the front of the queue. This is wrong." Accepting that nations had the right and duty to protect their own elderly, vulnerable, and health-care workers, he insisted that those same groups in other countries should not be forgotten. He continued, "I need to be blunt: The world is on the brink of a catastrophic moral failure, and the price of this failure will be paid with lives and livelihoods in the world's poorest countries."

COVAX vaccine did not reach Africa until 24 February 2021, when Ghana received a shipment; by then, more than two hundred million doses had already been given elsewhere. Distribution remained slow: by late September, thirty-nine of Africa's fifty-four countries had failed to meet the WHO goal of vaccinating 10 per cent of their people. COVAX pledges were not being fulfilled, and the next WHO goal of immunizing 40 per cent of Africans by the end of 2021 seemed beyond reach. Vaccines for Africa became a topic at the G20 summit in late October. By late December just seven African nations had reached the 40 per cent target, while average rate of vaccination for the continent was only 9 per cent.

Vaccine nationalism is also shortsighted. The world economy will not recover fully until the pandemic is controlled everywhere. The virus will keep mutating in unvaccinated regions only to return to those protectionist countries, naively believing in their herd immunity from a vaccine that might no longer work (chapter 12).

VACCINE DIPLOMACY

In contrast to vaccine nationalism and the western hoarding of vaccine, Russia and China were liberally applying a different rule: vaccine diplomacy. The two countries were using vaccine donations and sales to position themselves as generous, trustworthy, and powerful allies. They offered supplies to nations in difficulty, sometimes in quantities that surpassed their use at home. By early March 2021, China had promised five hundred million doses to forty-five low- and middle-income countries. Two weeks later, Sputnik V was licensed in forty-nine countries; whenever a slowdown occurred, as with Pfizer and AstraZeneca in Europe, Russia stood ready to help.

Willingness to accept certain vaccines was tempered by political views. On 22 March, an EU authority claimed that it had "absolutely no need" for Sputnik V, although the region was struggling to meet its vaccine targets and had been banning exports. Africa also proved to be a tough sell. In early February 2021, John Nkengasong, head of the Africa Centres for Disease Control and Prevention, expressed his opinion that token allocations should not be used to manipulate or curry political influence: "Africa will refuse to be that playing ground where we use COVID as a tool to manage relationships."

IN CANADA

Canadian provincial premiers and opposition parties blamed the federal government for the delays. The EU export decision placed Canada in a deeper quandary. Government leaders sought and received reassurances from their industrial counterparts that the reserved doses would not be affected. That led to more criticism of the EU for its arbitrary rules, favouring Canada over other states. As the CanSino product neared completion of phase-three trials, opposition parties criticized the federal government for having dealt with China in the first place, thereby postponing advance orders from other makers. But Canadian collaboration with that same Chinese entity had begun more than a decade earlier in the Harper era (Conservative), while closures and sales of Canadian vaccine-making facilities date from (and because of) the Mulroney era (also Conservative). The political derision was depressing and hypocritical, if not laughable, as if Justin Trudeau – feckless and clumsy though some perceive him to be – could be held personally responsible for the parlous state of global pharmaceutical trade.

The criticisms worsened when Canada announced in February 2021 that it would source vaccine from the COVAX stockpile, as it had arranged at the outset. On the one hand, Canadians saw large numbers of shots being given in both the UK and USA, and they complained over delays at home. On the other hand, they (sometimes the same "they") complained that it would be stealing from developing nations. Lost in an outraged din of politicians and journalists, was the fact that Canada had been the first G7 country to contribute a large amount to COVAX, established in April 2020. Intended to place all countries on a more equitable footing, the program considered the extent of the crisis and the state of local pharma manufacture to decide where to allocate resources. Canada's COVID-19 cases and deaths per capita were among the highest in the developed world (table 9.1), and its two-million-dose allocation had been planned all along. The United States did not participate at all in COVAX until after the Biden inauguration.

Even usually sensible physicians, exhausted by their work, began to pile on, demanding faster access to vaccines. A group of health-care workers, deep in the trenches, released an analysis on 21 January 2021, at Ontario's Science Advisory Table, the University of Toronto platform for pandemic data sharing. They advocated vaccine priority for all LTC residents, calculating how many more cases and deaths would result from the ongoing delays. Given current rates of infection, they argued, postponing vaccine coverage by just three weeks, from 21 January to 15 February, would result in approximately three hundred more deaths. The media had a field day.

But historians were perplexed. Surely current rates of LTC infection were totally unacceptable, and they were avoidable with proper sanitary measures. Certainly, some jurisdictions, like Kingston, Ontario, had seen little spread in LTC facilities, a fortunate but not inevitable outcome, achieved through careful monitoring by local public health. And why argue for 21 January 2021? Why not 21 January 2020? Or even 2019, when several million deaths might have been prevented globally had the world only been provided with a vaccine in advance.

In just a few short weeks, the amazing fact of having any vaccine at all had slipped from being a stellar achievement, a light at the end of the tunnel, to a deliberately blocked commodity. The frustrated, impatient, and very tired workers wanted someone to blame. Scapegoating again.

Worse, people craving release from the privations of control measures resented priority being given to others, especially if those others were reluctant to accept it. When planning began for doling out scarce vaccine, the premier of Manitoba realized that a large percentage of its initial share would be destined for the vulnerable Indigenous communities in his province. As if these people were not also Manitoban, he complained on 3 December 2020, well before the vaccines had arrived, that the recommendations were "unfair" and would "hurt Manitobans," putting them "at the back of the line." He claimed that Indigenous leaders agreed. But Indigenous leaders, together with many other citizens, were outraged and demanded an apology.

As the rollout continued and several million doses had been given around the world with only a few side effects, most vaccine hesitancy began to fade. And when the disappointing scarcity in supplies grew ever worse, some members of the initially hesitant groups began to crave vaccine, imagining yet another conspiracy that deprived them of protection.

Canada's first shots were given on 14 December 2020 from the Pfizer-BioNTech product. Because it had to be kept frozen at minus 70C, people in isolated communities and in the northern territories had agreed to wait for Moderna vaccine, which did not need extreme cold and could be transported more easily. It arrived in Canada on 24 December, one day after approval by Health Canada. The first shipment reached the Yukon and NWT on 28 December and Nunavut two days later. The military had been helping the rollout in advance by transporting freezers to the isolated communities. Priority would be given to elders and residents of long-term care.

Vaccine was accepted quickly when and where it appeared. Ontario was chastised for slowing vaccination between 25 and 28 December to give health-care workers a break. The workers themselves claimed that they would have happily kept going, and many others were volunteering to deliver doses round the clock.

THE VACCINE PROTESTS

The urgent demand for vaccine was not entirely built on selfishness and nationalism. The burden of disease, the strain on hospitals, the cancelling of so-called elective but life-saving procedures, and the rising death toll were reasons aplenty to pursue mass immunization. In addition, people were tired of lockdowns, hygiene rules, online schooling, and the social and psychological harm of isolation. In some places, they were deliberately and flagrantly disobeying the recommendations.

All along, groups around the world had voiced opposition to public-health measures. Wikipedia launched a page in April 2020, "Protests over responses to COVID-19 pandemic"; it kept growing to include dozens of countries on every continent. Some protests were motivated by rejection of the science, including tests, lockdowns, drugs, and the disease itself. Others objected for political reasons, financial stress, or simply from fatigue. In late January 2021, violent demonstrations against the rules became even more frequent. The Netherlands saw angry crowds battle police. Protestors in Los Angeles shut down a vaccination centre at Dodger Stadium. Belgium, with one of the highest COVID-19 death rates in Europe, witnessed massive demonstrations at railway stations and hundreds of arrests. Now the protesters targeted vaccines and vaccine mandates as well as the science and sanitary measures.

The right to protest may be fundamental, but large gatherings under lockdowns were illegal, having the potential to become superspreader events. Authorities established curfews, stiffer fines, and other punishments, but they could not monitor every situation.

At first, many protesters who objected to lockdowns believed that having had a vaccine meant that public-health rules were unnecessary, but, of course, the opposite was true. Even with vaccination, people had to continue precautions, although they might not fall ill. No one knew if vaccination would prevent spread of the virus. Gradually it became clear, with news out of Israel and Scotland, that vaccination might reduce the severity of illness and the need for hospitalization, but vaccinated people could still catch and spread the virus and suffer symptoms in what are called breakthrough infections. This message hit home in Canada early with 8 March 2021 reports of an outbreak among ten fully vaccinated residents of a British Columbia care home.

More disturbing was the rise and spread of variant strains from England, Brazil, Nigeria, South Africa, California, and India (chapter 9). Some appeared to be more contagious – making the continued precautions utterly essential and more stringent. In addition, the original vaccines had never been tested against those new strains. What other more dangerous variants would emerge if the pandemic was allowed to brew? The sooner vaccines were distributed everywhere, and the more tightly sanitary recommendations were maintained, the sooner the ordeal would be over: the pathogen would have less opportunity to make more evil mutations. If the vaccines proved ineffective against the new strains and if people were to let up their guard, COVID-19 would be stalking the earth for years to come.

Notwithstanding all the complaints, the uneven and seemingly slow rollout of vaccine, the world saw an interesting milestone on 3 February 2021: the number of vaccines given globally equalled the world's total of COVID-19 cases. Various news sites joined the health agencies to feature daily updates on each nation's progress with immunization. It was a race against the virus, against time, and, sadly, against each other.

Also in early February 2021, the *Lancet* finally published the long-awaited trials on Russia's Sputnik V vaccine. Based on a study of more than twenty thousand mostly white, mostly male subjects vaccinated in Moscow between September and November 2020, it confirmed more than 95 per cent effectiveness. This news was greeted as good for health care and as a big political win for Vladimir Putin. Beyond economics, vaccine wars were intensely political.

By mid-February 2021, Israel still led the pack with 70 per cent of its population vaccinated, approaching the much-touted herd immunity. UK was at 21 per cent with a dramatically falling case rate. The United States was at 13 per cent with an impressive 1.6 million shots being given each day, unevenly distributed across states, surprisingly led by West Virginia and North Dakota. Canada's proportion was much lower at 3.2 per cent vaccinated. But few of the grumbling people understood that, at the same time, Canada's rate was just under the EU average of 4.4 per cent and well beyond several EU members, including the Netherlands. Moreover, Canada had moved ahead of countries that manufacture vaccines, such as China, Russia, and India, and most other nations of Middle East, Asia, Africa, and Latin America. Nevertheless, citizens and especially politicians did not stop whining.

In the weeks following the Biden inauguration, Canadians felt let down and angry that the doses ordered from Pfizer had to be supplied from its now closed Belgian plant, when there was stock just across the border. Toronto infectious disease expert, Dr Isaac Bogach, complained to Global News on 11 February: "Our best friend and neighbour is producing vaccinations so close to home you could shoot a hockey puck and hit Ontario from the factory in Kalamazoo (Mich.). But they're not sending any vaccines our way. So, we have to get them from Europe." Bogach had given yeoman service to the media in patiently explaining the pandemic and its management; his frustration was poignant.

OXFORD-ASTRAZENECA WOES

In February amid global craving, came some disappointing news. South Africa announced that it was suspending the rollout of a million doses of Oxford-AstraZeneca vaccine, which it had just received, because it was deemed ineffective against the new strain (now called *beta* variant). What would happen to those millions of doses? Would anybody want them now?

The vaccines nearing completion later in the pandemic, such as Johnson & Johnson and Novavax, could be tested against newer variants; results suggested that they would have reasonable effectiveness. Manufacturers of the novel mRNA vaccines (Pfizer and Moderna) promised that they could modify their products for variants, although variants had not been part of the original trials. People began lobbying for the right to choose a preferred vaccine.

In early March 2021, Italy was ravaged by its fourth wave with three hundred or more deaths daily. It decided to use the new EU protectionist rule to block a shipment of 250,000 doses of AstraZeneca vaccine made in a factory near Rome and bound for Australia where no one had died for weeks. Australia appealed as a matter of principle, but France and other EU countries revealed they too were considering using the rule. On 24 March, the EU revealed more draft legislation that would curb all vaccine exports, an extraordinary intrusion into the commercial workings of companies hosted in their region, many of which were not EU owned. The ban didn't materialize, but the threats continued, while the earlier EU mechanism for transparency and authorization of exports was strengthened and extended into 2022.

But AstraZeneca's troubles were not over. In the middle of the export controversy, on 11 March, Denmark suspended use of the same vaccine because of concerns over the possible side effect of blood clots. More than a dozen countries quickly followed suit. Ironically, the list included Italy that had blocked the shipment to Australia shortly before.

Clots had occurred in fewer than forty of the twenty million people who had received the AstraZeneca shot. This rate was no higher than what was anticipated in the unvaccinated population. It took another week, until 18 March, for cases to be investigated and the vaccine declared safe. But scientists kept working on the problem because the handful of affected people had been young and otherwise healthy. On 21 March, German scientists found an explanation in a recognized and treatable syndrome that operates through platelets.

In the meantime, skepticism about AstraZeneca mounted. The UK prime minister, Boris Johnson, quipped, "So the Oxford jab is safe, and

Figure 11.1 | Editorial cartoon by Bruce MacKinnon, *Chronicle Herald*, Halifax, 21 March 2021

the Pfizer jab is safe. The thing that isn't safe is catching COVID." He rolled up his sleeve and took his AstraZeneca shot, as did Jean Castex, prime minister of France.

But AstraZeneca was rapidly becoming the one vaccine that nobody wanted (figure 11.1). Many factors combined to feed a targeted hesitancy: middling effectiveness; the fact that trials had not included the elderly; failure with the South African *beta* variant; and now possible side effects. Little wonder then, Canadians thought, when on 18 March 2021, the very day when the European Medicines Agency confirmed its safety, the United States finally offered to loan Canada and Mexico some its jealously hoarded vaccines – vaccine that would soon be outdated, vaccine it was not using, vaccine it had yet to approve: AstraZeneca, of course.

An American trial on AstraZeneca, reported in March 2021, did include the elderly and suggested good safety and higher efficacy than original studies on younger people; however, the United States has still not licensed this vaccine. As of July 2021, it accepted AstraZeneca as vaccination proof when it had been given elsewhere. In the meantime, declining caseloads there and in Britain, already seemed to confirm the protective value of vaccines with only 38 and 45 per cent of citizens, respectively, having had at least one dose. Was it vaccine? Or, was it the weather?

THE ROCKY ROLLOUT: CARROTS AND STICKS

Canada finally began receiving more reliable supplies of vaccine in late March and April 2021. Regions began acting on plans for rolling out shots, first to Indigenous groups, the most vulnerable, and health-care workers, then to cohorts by declining age, reaching children twelve years and older by June. Hoping to allay fears still swirling around AstraZeneca and not wanting to jump the queue, Prime Minster Trudeau received his first dose of that unpopular vaccine in late April; his second shot was Moderna, in July.

Most people eagerly booked appointments. Volunteer workers in the vaccination centres were charmed by the happy atmosphere – a stark contrast to the gloomy pall that had hung over life for more than a year. By June 2021, Canada had recovered from its slow start and was soon leading the world in immunization – even exceeding rates in countries that manufacture vaccines, such as Switzerland, the UK, and the US (figure 11.2). The noisy complaints over delay and incompetence ceased. Instead, Canada was being criticized for having booked so many vaccines for itself when the rest of the world was in want.

Country after country learned existentially that the vaccines were effective. By autumn 2021, severe COVID-19 had become a disease of the unvaccinated (figure 11.3). But even with 80 per cent of the population vaccinated, outbreaks dominated by the unvaccinated continued to challenge health-care facilities. Media reported on individual anti-vaxxers and COVID-19 deniers, who had barely survived infection, urging their fellow citizens to take the shot. Recognizing the possibility of side effects, Canada established its first Vaccine Injury Support program in June 2021; six months later, it had received just four hundred applications only five of which were approved. Policy makers tried many strategies to incentivize vaccination in regions that needed it most.

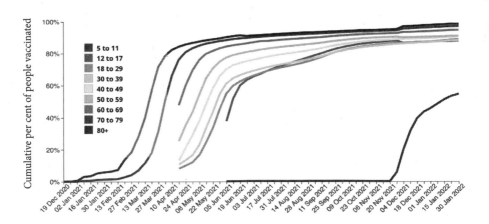

Figure 11.2 | Percentage of eligible people vaccinated in Canada, by age group, in cohorts of declining age, December 2020 to January 2022 (Source: Health Canada, https://health-infobase.canada.ca/covid-19/vaccination-coverage/, accessed 17 January 2022.)

American jurisdictions offered prizes for taking the shot: cash, free beer or wine, concert tickets, groceries, and lottery tickets for an odd variety of desirables: hunting or fishing licences (Arkansas, Maine, Minnesota, and West Virginia); driving a racing car on a high-speed track (Alabama): and university scholarships (Delaware). Alberta ran three lotteries during its famous "Open for Summer" 2021 program, awarding a million dollars on three occasions to citizens over age eighteen who were vaccinated. It also randomly distributed vouchers for air and train travel, sports events, and the Calgary Stampede. But by December 2021, analysis had shown that these incentives do not work. Vaccine-hesitant people are not nudged by gifts; for some they backfired, acting as a disincentive by implying that there was something wrong with vaccines. Only mandates that affected travel and employment had any hope of encouraging the hesitant.

The positive effect of vaccines soon led many countries to recommend vaccine certificates for activities, mandates for employment and schooling, and passports for travel. From July 2021, vaccinated citizens of the EU could obtain a Green Pass allowing travel across borders. Unvaccinated citizens and tourists everywhere would be obliged to have

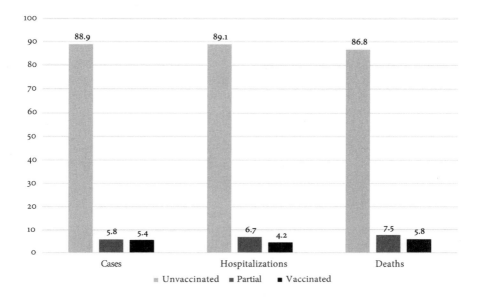

Figure 11.3 | Distribution of confirmed COVID-19 cases reported in Canada by vaccination status as of 2 October 2021 (per cent) (Source: Health Canada, https://health-infobase.canada.ca/covid-19/epidemiological-summary-covid-19-cases.html#a9, accessed 25 October.)

frequent tests. In October 2021, France encouraged vaccination among its holdouts by charging for the previously free tests. Proof of immunization would be essential for work in health care, government, and education, and it would permit access to previously restricted activities, like theatre, concerts, and dining out.

Through August and September 2021, Canadian provinces began announcing rules and deadlines for vaccine certificates. The carrot for skeptics would be the fact that vaccines could shorten and personalize the lockdowns if not eliminate them.

The postelection federal government announced a forthcoming vaccine passport for travel outside the country. It coincided with the news that, after eighteen months, the land border into the United States would reopen for Canadians and Mexicans in early November 2021. US citizens had been free to enter Canada since August. Canadians had been free to fly (not drive) into the southern neighbour; the opposite courtesy did not apply.

Eventually, in a strange reversal, the noisy protests of early 2021 switched from demands for more vaccine to refusing immunization altogether, all the while continuing to object to lockdowns. As anticipated, breakthrough infections had occurred. News in September 2021 of the side effect of myocarditis (heart inflammation) in young males who had received mRNA vaccines increased anxiety, although the effect was transient, and the risk of the same problem was far greater in full blown COVID-19. Unaware of overall statistics, protestors used the anecdotal facts of rare side effects and breakthrough infections to insist that vaccines were useless and dangerous. They also lied: traffic in false certificates and fake medical exemptions emerged.

Appalling demonstrations took place outside hospitals, blocking entry for sick people, ambulances, women in labour, and the willing workers. Rallies at Canadian hospitals increased before and after the 20 September 2021 election, until they were made illegal in December 2021. Some demonstrations had been organized by dissenting hospital staffers, refusing to accept vaccine for employment. Various labour unions also objected to vaccine mandates. Incredibly the nurses' union of British Columbia disagreed with the mandates (not with vaccine), arguing that they would aggravate staff shortages; a backlash from members coincided with the sudden resignation of the union president. Public-health physician Bonnie Henry, once a media darling, received death threats, demands for her resignation, and cries of "Lock-Her-Up!" that bore unmistakeable similarity to the chants of Trump supporters against Hillary Clinton. Nevertheless, province-by-province health-care workers had to take the jab or be suspended without pay. In early October, despite staffing shortages, hundreds of unvaccinated hospital employees were suddenly out of work. Most eventually accepted the shot.

Few protestors recognized that their vision of personal freedom included the right to make others sick. Some invoked their deity, claiming that unalienable liberty had been ordained in the Bible, although God and freedom had had little to do with each other until slave-owner Thomas Jefferson linked them in the American Declaration of Independence. Anti-vax placards proclaimed, "My body, my choice!" Ironically, this rhetoric had been purloined from the women's movement demanding access to abortion; yet anti-vaxxers included many opponents of abortion and women's rights (figure 11.4). Not only did they reject the science, the favourable statistics, and the policies, they had little notion of the long history of well-tolerated vaccine mandates.

Figure 11.4 | Editorial cartoon by Adam Zyglis, *Buffalo News*, 18 June 2021
(Courtesy of Cagle Cartoons.)

They believed that, somehow, these recommendations were new and unprecedented. They were not.

Religious leaders from most faiths and doctors pointed out that valid exemptions were exceedingly rare. As the rollout moved on, studies revealed that vaccine protection waned through time. With the variants and breakthrough infections, some experts, including the US FDA and some Canadian provinces, began advocating third shots, or boosters, especially for vulnerable people and the elderly.

A new rush on vaccine took place. In October 2021, Pfizer announced positive results on trials for children, drawing mixed reactions from parents – relief and skepticism. Doses became available for Canadian children as young as age five in late November. At first uptake was rapid, but it stalled with roughly 50 per cent of eligible preteens immunized by January 2022. Parents hesitated more to vaccinate their children than themselves, believing that the disease was mostly mild in kids and too many unknowns still swirled around vaccines – despite the nine billion doses given globally. Newfoundland led the country in child vaccination citing its culture of community spirit, respect for the elderly, and indelible memories of the past horrors of infectious disease.

When Singapore decided to charge unvaccinated people for their medical expenses, some Canadians opined that we should do the same, especially since the new drugs were so very costly. Along similar lines, Quebec attracted international attention in January 2022 when it briefly considered a tax on the unvaccinated, to cover the heavy expenses that they caused.

Psychiatrists pleaded for better understanding, more respect, and compassion in dealing with the hesitant; some fears were plausible, they claimed, if unrealistic, deriving from historical blunders and imaginative science. Purveyors of the infodemic got involved too, converting many to their views. Among the most visible were physicians who, like psychiatrists, worried about the harms of isolation and masking. and who also objected to vaccines and, especially, to the mandates; some were providing unjustified exemption certificates, while others attempted to sue government. Dissenting doctors in British Columbia, Alberta, Ontario and elsewhere were cautioned or stripped of their practice licences for spreading misinformation and what was deemed unprofessional behavior. As if medical advice was solely a matter of personal opinion, some ethicists began to question whether or not physicians were entitled to free speech. Suspicion of authority was rampant, and citizens began to criticize what they perceived as the discriminatory ethics of test certificates and proposed vaccine passports for work, entertainment, or travel. Aware that mandating vaccines could backfire, leaders had insisted, all along that they were for "everyone who wants one"; no one would be forced.

On one hand, the idea of vaccine passports and QR codes on smart phones offered a kind of simplicity and a return to liberty that had been curbed for too long. On the other, anti-vax fabrications contended that vaccines would inject particles, microchips, that could be surveilled and controlled by evil forces anywhere, even from outer space. When mandates limited work or play, then the hesitant viewed them as coercive as well as dangerous; they claimed that they had indeed been forced, despite the promises of politicians.

The debate raged between those who believed that increased surveillance was the only logical way forward to protect a vulnerable collective – and those who saw surveillance as a prison without walls or the eye of Big Brother.

The rocky rollout simply emphasized, once again, the lack of basic preparedness. Even as the developed world tilts toward quixotic herd immunity, it still has much to learn about transporting vaccine, establishing sites for administration, finding staff, booking appointments, encouraging acceptance, and managing protests. On this background, arose the growing probability that vaccinating everyone would not be enough. We might need yearly boosters, and vaccinated people can harbour and spread the virus. Given time and continued disease, new variants will arise resistant to extant vaccines. Therefore, having the shot did not absolve people from obeying the old rules of public hygiene. The vaccine was not a cure.

COLONIALITY OF PUBLIC HEALTH

For the past two decades, social scientists have written of coloniality in a postcolonial world. It is a legacy that entails structural and cultural manifestations of paternalism, exploitation, marginalization, pathologization, and saviourism. Some contend that epidemiology and public health are rife with it.

Whether or not one accepts the idea of this process, there was no denying the shocking discrepancy in global immunization rates by late 2021. As Canada approached 80 per cent of its total population being vaccinated, statistics from Africa, Latin America, and Haiti remained depressingly low, at far less than 10 per cent. Mostly the figures reflected a lack of vaccine. These statistics ought to bring a sense of shame to wealthier nations. The message that "we are not safe until everyone is safe" has yet to be learned. Struggle to manage the pandemic will continue.

Nevertheless, developing countries can be impressively capable of conducting immunization campaigns if they possess vaccine. When India provided supply, Bhutan succeeded in vaccinating its entire population in a single week. Confronted with crippling American sanctions, Cuba turned its own vaccine industry to COVID-19 and exceeded the immunization rate of the rest of the world using its homegrown vaccines and reaching children as young as age two. The effectiveness is yet to be determined, but the resourcefulness cannot be denied.

While developed nations moved on to boosters, the COVAX plan was starved for vaccine. Furthermore, the doses available in Africa through COVAX had mostly been Covishield, made by the Serum Institute of India, licensed by AstraZeneca – a product that would not be recognized by emerging vaccine passports. Wealthier countries pursued ever more product through separate contracts with firms. Israel began recommending fourth doses. WHO Director Tedros repeatedly begged developed nations to delay boosters and child immunization to share vaccine with the rest of the world. His WHO colleague Mike Ryan described third shot campaigns as unethical: "handing out extra life jackets to people who already have life jackets, while leaving other people to drown."

A new wave was on the horizon that would challenge progress, assumptions, and predictions everywhere.

12

Omicron and the Origins

In late November 2021, just as the vaccine rollout reached major milestones in the developed world, dismaying news came out of South Africa. Scientists had found yet another variant of concern that appeared to be highly contagious. Was it more lethal too? Could it be worse than *delta*? After nearly two years of complying with hygiene rules and enduring intermittent lockdowns, people heard this news with skepticism, fatigue, and fear. Several countries, including the United States and Canada, promptly banned flights from South Africa and its neighbours, Namibia, Zimbabwe, Botswana, Lesotho, Eswatini, and Mozambique – some of which had seen no cases at all.

Omicron became the name of the new variant (B.1.1.529). The Greek letters *nu* and *xi*, which precede it, were skipped to avoid giving offence: *nu* was a popular Yiddish expression, and *xi*, the name of the Chinese leader. Once again, however, scapegoating persisted. Distinguished scientists in South Africa had identified the variant and alerted the world. For their diligent work and globally responsible actions, however, their region endured cancelled flights, economic hardship, and trade restrictions – although the variant had probably originated elsewhere.

Angry commentators, including WHO officials, pointed out that travel bans were harmful and ineffective. Indeed, the variant was later shown to have already been circulating in Europe well before it had been identified. Bans, they complained, were a form of "travel apartheid": countries merely succeeded in stranding their own citizens abroad and in punishing the messenger – an African messenger. The bans were lifted three weeks later in mid-December.

The highly contagious *omicron* variant quickly caused global case numbers to explode, reaching unprecedented levels and dwarfing the

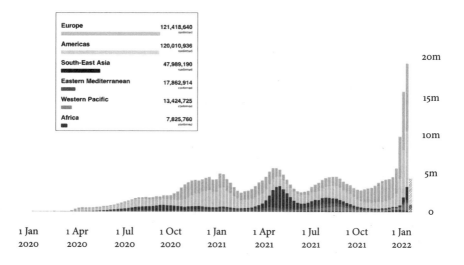

Figure 12.1 | Weekly global COVID-19 cases by region, January 2020 to January 2022 (Source: World Health Organization, https://covid19.who.int, 19 January 2022.)

peaks of the other devastating waves (figure 12.1; compare with figure 9.2). It made a mockery of the complacency that had settled on the vaccinated sectors of the world, just about to dispense with restrictions as a reward for having complied with recommendations. It confirmed prophetic warnings: the pandemic would not be stopped until the entire world had access to vaccines.

Omicron also became a soldier recruited to both sides of the vaccine wars. On one hand, vaccines soon proved less effective in preventing it than its predecessors; the surging cases included breakthroughs in the fully vaccinated. The anti-vax lobby construed this fact as more evidence that vaccines do not work. On the other hand, *omicron* ramped up existing calls for booster shots because severe illness and death were less likely with greater immunization coverage.

By mid-December, all European and North American countries had experienced an almost vertical rise in cases soon after the arrival of *omicron*. Eventually hospitalizations and deaths rose too, but to a lesser extent than previously. The personal stigma of testing positive declined.

Soon everyone knew someone who was sick, even if they had been following rules and taken boosters; unvaccinated children seemed more vulnerable and brought the disease into homes. Despite its dizzying caseload, the UK at first seemed to have a lower case-fatality rate than other places. Was this an effect of improved behaviour? Or, was it an effect of the Oxford-AstraZeneca vaccine, so widely used in that country? France had tried to block travel from UK earlier in the fall 2021, prompted by its higher caseload. But soon its own count soared to a half million cases daily, a European high, surpassed only by Denmark, which may have been counting more accurately. The surge made France's proposed restrictions against the UK seem petty and sanctimonious, as well as useless.

"Flatten the curve" vanished. This was not a curve, it was a spike, a towering skyscraper. Over two years, Australia had seen impressively few cases and had recently reopened for long-stranded citizens to return. But with *omiron* its accumulated case numbers tripled in less than a month. Unaccustomed to social-distancing measures, Australians had trouble believing the warnings and following the rules; its case fatality rate soon exceeded previous levels. Even New Zealand began seeing a surge that prompted tighter restrictions; its prime minister postponed her own wedding. At the opposite extreme, Mexico's test positivity rate grew and its already high case fatality rate rose again (table 9.1); its leader caught COVID-19 for the second time: wearing no mask, he called it, "a little cold."

The learning continued. Not only was *omicron* more contagious, it had a shorter incubation period than other variants, and somewhat different symptoms. Where fever, cough, difficulty breathing, and loss of the sense of taste or smell had been the most frequent symptoms of earlier variants; *omicron* sufferers reported upper airway symptoms of runny nose, headache, sore throat, fatigue, and sneezing. More children seemed to be affected. The South African experience hinted that it might not be as lethal and that, perhaps, it would simply wane within a month. But South Africans were younger, and given the recent two years, they might possess considerable natural immunity. At least two weeks should pass before the true severity could be determined.

Omicron seems to have entered Canada via eastern Ontario. It was first found in Ottawa in late November in two people who had travelled from Nigeria. Within days it appeared everywhere in the country, and by mid-December, it had replaced *delta* as the dominant strain. The caseload made earlier waves appear to shrink (figure 12.2; compare with figure 9.3). *Omicron* might be relatively mild, but with astronomical

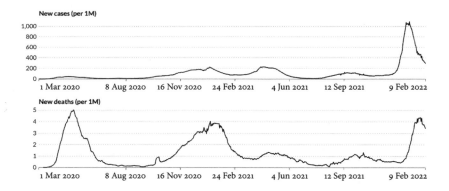

Figure 12.2 | Incidence of COVID-19 cases and deaths in Canada, March 2020 to February 2022 (Source: Our World in Data, https://ourworldindata.org/covid-cases, accessed 26 January 2022.)

numbers of infections, even a tiny proportion of the infected falling severely ill was more than sufficient to overwhelm hospitals. Deaths climbed to match – although not dwarf – earlier peaks.

Lockdowns, curfews, and online school were reinstated variably across the country. Just as the other financial supports had been coming to an end, the federal government quickly introduced a new program called the Canada Worker Lockdown Benefit hoping to help allay stress and encourage those feeling sick to simply stay home. For that had become the rule: "if you have symptoms, assume it is *omicron*, and stay home." From once having the best figures in the country, our region of Kingston went (briefly) to the worst, with almost three hundred new cases daily for a population of two hundred thousand. Our total mortality, which had remained low with six deaths in two years, quadrupled in a month. Non-COVID operations and admissions were delayed yet again.

The new variant disrupted and mocked our previous pandemic practices that had become almost routine. So many people became contacts of cases that it was impossible to reach or test them all properly. At first, however, a negative PCR clearance test was needed to leave isolation. Test centres frantically tried to meet demand, but available appointments grew scarce, provoking anger and frustration. Public-health nurses encountered ever more abuse. In some places, the test positivity rate soared to almost 30 per cent, indicating many more uncounted infections. Recall

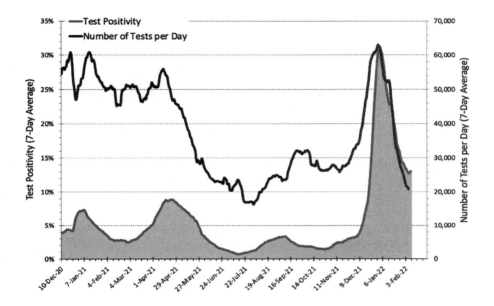

Figure 12.3 | Test positivity and number of COVID-19 tests in Ontario, December 2020 to February 2022 (Source: Science Table, https://covid19-sciencetable.ca/ontario-dashboard/, 23 January 2022.)

from chapter 5, a positivity rate above 1 per cent means measures are incomplete; at 30 per cent numbers are probably meaningless (figure 12.3). Workers were exhausted; laboratories could not keep up. The rules had to change.

On 31 December 2021, comprehensive contact tracing was abandoned in several provinces because callers could not reach them all. Our unit contacted only people who were unaware of exposures: for example, patients who had been in clinics or ambulances. Simultaneously, the gold-standard PCR tests, which had been essential for ending isolation, were restricted to priority cases, such as health-care workers. People with suspicious symptoms – tested or not – were urged to isolate and notify their friends: do-it-yourself contact tracing. Yet quarantining and isolation remained vital to control, and some form of measurement was essential for assessing management.

The chronic shortage of health-care workers worsened. Burnout, illness, and death had diminished the already inadequate staffing numbers. With the *omicron* surge infecting caregivers and forcing their exposed colleagues into isolation, hospitals began allowing infected caregivers to continue working, starting with Quebec on 28 December 2021. Without their continued support, care would collapse. Of course, they would wear full PPE and be assigned where transmission would be of lesser danger. Some nurses and members of the public objected loudly to the risky decision. Nevertheless, within a month, other Canadian, American, and French hospitals and LTC facilities were introducing the same procedures – procedures that would have horrified public-health workers two years earlier.

By 8 January 2022, the Health Canada graph in figure 11.3 still showed an advantage with vaccines nationally, but the percentage of unvaccinated cases had fallen to 56.6 per cent and the vaccinated cases had risen to 35.7 per cent. In Ontario, things were worse: around 25 December 2021 the incidence of cases among vaccinated people with COVID-19 began to exceed the unvaccinated cases, although absolute numbers of patients in ICU were roughly equal. Since many more people were vaccinated than not, those equal ICU numbers came from pools of vastly different sizes. The risk of severe disease, requiring oxygen, or death could be up to ten or twenty times worse in the unvaccinated; an American study suggested it was forty times worse. Charting incidence – that is, rates per capita – was more meaningful, but it was difficult to grasp. Everybody knew someone who had been infected, and often enough it was someone who had been vaccinated.

The previously disparaged rapid antigen tests were recommended instead; however, they were in short supply everywhere. As the winter holiday began, Ontario schoolchildren were each given five tests to use in case of symptoms and before the planned return to class. Tests were distributed in grocery and liquor stores, but there were never enough. Some were offered online at great expense and of dubious quality. Officials warned that their accuracy was unreliable in the absence of symptoms; however, they were better than nothing, and families used them for reassurance before visiting elders during the holidays.

Without comprehensive testing, public-health officials had to turn to other indicators to track the severity of the pandemic. They looked to hospitalizations and ICU admissions as indicators of spread. With widespread infection, however, some in-hospital patients who tested positive had been admitted for reasons other than COVID-19; it was difficult to

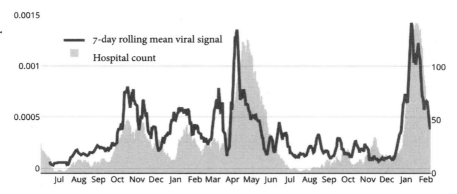

*The wastewater signal usually rose ahead of the clinical diagnosis. Note that the earlier decline in case numbers during the omicron wave is apparent, not real, owing to the restricted testing while the wastewater signal continued to show viral presence.

Figure 12.4 | 7-Day mean COVID-19 wastewater viral signal, Ottawa, July 2020 to February 2022 (Source: Ottawa, COVID-19, at https://613covid.ca/wastewater/, 23 January 2022.)

distinguish these incidental cases from others. There had to be other ways of knowing if the *omicron* wave was getting better or worse.

At this point, wastewater testing grew in importance. Some jurisdictions, like Ottawa, had been checking sewage all along, finding that the results reliably reflected the successive waves identified with PCR testing of individuals (figure 12.4). Now wastewater tests became a sought-after indicator of the extent of infection within a community. An Australian study indicated that they could be used on airplane sewage to warn of arriving infections even when travellers' tests are negative. But having originated in environmental science, wastewater tests have not been widely integrated into public health; their use is still patchy.

MORE LOCKDOWNS, FATIGUE, AND PROTESTS

Meanwhile China maintained its zero-tolerance policy. With its in-person Winter Olympics set to begin on 4 February 2022, it imposed an indefinite lockdown on the thirteen million people of Xi'an in late

December 2021. Two more cities entered lockdown in response to *delta*-variant cases, Yuzhou (1.1M), and Anyang (5.5M). China's first *omicron* cases, reported in Tianjin in mid-January, were attributed to a package mailed from Toronto via the United States and Hong Kong. Experts denounced the claim as a "ludicrous" manifestation of tensions between the two nations. By 24 January, the three-week lockdown on Xi'an was lifted, and the national caseload had dropped from one hundred daily to eighteen; China then implemented mass testing of two million people in the Beijing district of its soon-to-open Winter Olympics. Athletes and other visitors would undergo daily testing, although, at first, Chinese officials had selected highly sensitive tests that picked up those who had experienced COVID-19 and recovered, but they adopted more reasonable tests in response to the complaints. The strict measures were widely publicized, and some criticized them as a recipe for disaster in that vast country where its own vaccines do not work against *omicron*. By the end of the games on 20 February, just over 400 infections had been detected and contained, suggesting that others wishing to host big events could derive lessons from that experience. Nevertheless, just over a week later, China posted its highest case numbers thus far, although they were far below the US *omicron* peak in a population less than a quarter the size. But few nations can emulate China's vigilance, for lack of human and material resources and, especially, for lack of acceptance.

Many people in the West simply would not put up with the ongoing measures and they were angry that vaccines had failed to protect them. The stress of repeated lockdowns, curfews, closed businesses, online learning, and virtual work began to generate pushback from weary citizens everywhere on an even greater scale than in the past. No scientist had promised that vaccines would be a panacea, but the rising caseloads generated dismay. Many, especially those suspicious of authority, felt that they had been lied to yet again, just as they viewed vaccine mandates as betrayal and a form of coercion, which, leaders had promised, would not happen.

Some engaged in wishful thinking: the pandemic was dying with a spectacular blowout before shriveling into herd immunity: this would be the end; the pandemic will morph into an endemic annoyance. Others accepted vaccines but opposed mandates, mistakenly claiming that they were unprecedented. The objections were barnacled with far-right notions of freedom, liberty, and conspiracy theories over pandemic management, treatment, and even its existence: a motley crew of strange bedfellows. A Toronto psychiatrist, possibly in a misguided attempt to raise understanding for anti-vaxxers, penned a lengthy *Globe* op-ed, alleging lies and censorship, and distorting the past; his essay prompted exasperated explanations from people who toil in the COVID trenches and an editorial that leaned toward mea culpa from the publisher.

Various political entities quickly joined the ruckus, profiting from media coverage. Misinformation was rife. Death threats multiplied. Crowds menaced doctors and elected politicians outside their homes. Violence erupted in Belgium, Germany, the UK, and on the French islands of Guadeloupe and St Pierre and Miquelon. Sad irony reigned: banners and badges characterized the government and public-health leaders, who were desperately trying to save lives, as Nazis and fascists – an offensive charge, particularly wounding to those who had family members murdered in the Holocaust.

A massive flood of Freedom March protests was organized globally on the weekend of 22–23 January 2022. Tight crowds of unmasked protesters surged through the Washington Mall, and the streets of great cities – Prague, Paris, Vancouver, Montreal, and Ottawa – and of small cities too, such as Moncton and Kingston, blocking traffic and chanting all manner of hatred.

Politicians condemned the marches and the personal attacks but feared their power. When hospitalizations had plateaued at peak numbers and the death rate was as high as it had ever been, Denmark decided to lift restrictions for vaccinated people wearing masks. Many countries in Europe planned to follow its lead. Days later, Danish scientists announced yet another variant of *omicron* (BA.2) – nicknamed "stealth omicron" – because it escapes detection on existing tests and already accounted for half that country's caseload.

Similarly, with hospitalizations at a peak, deaths still rising, and Prince Edward Island suffering its first severe wave, the premiers of Ontario and Quebec promised more re-openings by 31 January 2022. Both were facing elections later in the year. Sounding like their Alberta counterpart of the previous summer, they claimed that we would just have to live with COVID-19 and learn to accept a few deaths. No one in Europe, Canada, or the United States could aspire to the kind of control that was operating in China.

On 25 January 2022, Pfizer-BioNTech announced the launch of clinical trials on its new *omicron*-specific vaccine – not to ecstatic joy, as before – but to a weary sense of Groundhog Day – here we go again! Calls for broad-spectrum vaccines that could deal with any variant grew louder, with reminders that sanitary measures should not be abandoned.

On the Freedom March weekend, convoys of truckers set out from British Columbia in the west, and Nova Scotia in the east, rolling to Ottawa on a deliberate traffic slowdown to dispute the 15 January implementation of vaccine mandates for border crossing. The action was promptly denounced by the Canadian Trucking Alliance, 90 per cent of whose members are vaccinated. But media covered the protestors, inadvertently (or deliberately) promoting their agenda. Supported by a crowd-funding source that gathered millions of dollars, with sizeable amounts from American donors, the truckers waved Canadian and US flags and banners featuring Donald Trump or likening the prime minister to Hitler. Reaching the capital on 29–30 January, they shut down the city core, desecrated monuments, stole from food banks, set loud horns blaring, and hurled expletives and fireworks. Other angry groups blocked border crossings at Windsor and in Manitoba. Trying to balance the right to demonstrate against protecting public safety, police were overwhelmed, and businesses were forced to close. On 31 January, Prime Minster Trudeau tested positive for COVID-19; he refused to meet with the truckers, although several conservative politicians conspicuously supported the protesters. The convoy stayed for three weeks. Eventually on 14 February 2022, the prime minister invoked the *Emergencies Act* (EA), which stood in effect till 23 February. Bank accounts of the leaders were frozen, and many arrests followed. Meanwhile, mirroring the Ottawa debacle, convoy protests arose in several other countries. The controversial decision to invoke the EA was subjected to an obligatory inquiry that began in late April 2022.

On 24 February, just as the chaos of the Canadian trucker convoy began to resolve, Vladimir Putin broke all his promises and invaded Ukraine. The wanton destruction, the killing of civilians, the evidence of war crimes, and the threat of nuclear weapons displaced the pandemic from the news. Nevertheless, some observers were quick to point out that the conditions of conflict have always been ripe for the spread of disease: like truth, health measures are an early casualty of war. The invasion began as Ukraine was in the midst of an *omicron* surge, case counting vanished, and millions of people fled into other countries taking their children and their germs with them. The pandemic is not over.

Historians often explain that pandemics do not end at a point in time: endings can be biological (virus), medical (disease and controls), and social (public acceptance); the endings do not necessarily coincide. The convoys and the premature re-openings, in the face of ongoing *omicron* infection, proclaim the wisdom of that observation. Some people had simply decided that the pandemic was over and behaved accordingly, even as more people were falling ill. Anthony Fauci seemed to endorse the social and medical views on 27 April 2022 when he declared that the United States was out of its "full-blown, explosive pandemic phase," although he cautioned that the world is still in a pandemic and conceded that new variants will likely arise.

It is no accident that the highly contagious *omicron* variant originated where vaccines are scarce. Fostered by vaccine greediness elsewhere, it consolidated the message – unpopular though it may be – that no one is safe until everyone is safe. At least, most vaccines seemed to protect against severe disease – for the time being. But what about future variants and future viruses? Because surely more will come. Could some answers be found in the past?

ROLLING BACK THE ONSET: MOLECULAR ARCHEOLOGY

During the struggle with SARS, some pundits expressed great anger with doctors in China's Guangdong province for failing to warn the world of the emerging disease. At the time, I wondered if I would be capable of recognizing a completely new disease, especially one that displayed the common symptoms of fever, cough, and respiratory difficulties – symptoms that physicians encounter daily. Everyone knows that some colds and flus seem far worse than others and are more plentiful in some seasons than others. Actively suppressing information about a new pathogen is unacceptable. But first, one must have confidence that the pathogen is new. Being able to make that observation might require the tiny handful of initial sufferers to go to the same clinic. I marvelled at the urge to blame and find a culprit.

Once tests for COVID-19 became widely available, scientists began looking retrospectively for the earliest evidence of the new virus. The exercise exposes how the virus spreads and might provide guidance for future policies. To accomplish that task, they needed several things. First, they had to find people who had been ill or died with similar symptoms before the official date of recognition. They also needed reliable, datable samples from those people taken when they were sick, and they needed new antibody tests from survivors. To explore specific pathways, they could trace the geographic location of various viral strains. Another approach would apply the new COVID-19 tests to historical samples of wastewater because the virus was excreted in feces. But for how long are sewage samples stored, and where?

The evidence began appearing in March 2020, when Chinese news reported that its earliest confirmed case was a fifty-five-year-old man from Hubei province who fell ill on 17 November 2019. Before the end of November, ten other people – five men; five women, aged thirty-nine to seventy-nine – were believed to have had a disease like COVID-19. By 31 December 2019 and the earliest warnings, the number of presumably infected Chinese people had risen to 266, although most of those illnesses had not been recognized at the time as representing a new disease.

Over the course of 2020, other reports trickled in from different countries with evidence that SARS-COV-2 was present outside China before the Wuhan outbreaks became apparent and health measures began. Here are a few examples:

- On 12 April 2020, the *New York Times* blamed an in-person international conference held in Boston on 26–27 February for having been a superspreader event that disseminated infection to one hundred people. By late 2020, analysis showed that the specific strain of virus at that conference had probably come from France and was implicated in up to three hundred thousand cases worldwide. Ironically the conference was hosted by Biogen, a company involved in making monoclonal antibodies.
- On 21 April 2020, the US announced that two individuals, not previously known to have had COVID-19, had actually died of the disease in California on 6 and 17 February.
- On 5 May, coronavirus was confirmed in a forty-three-year-old man admitted to a Paris hospital on 27 December. He recovered. His family members had also been ill. He had not travelled abroad.
- A week later, Brazil reported that a man who had died around 25 January had succumbed to the virus, although that country's first recognized case did not appear until 25 February.
- By June 2020, wastewater studies placed the virus in Italy on 18 December 2019.
- Later in June, studies from Spain identified the virus in Barcelona wastewater in mid-January, at least forty days before its first official case. Another sample from 12 March 2019 [sic] also tested positive, raising doubts about its accuracy and demands for retesting.
- Also in June, analysis of symptoms of patients seen from December 2019 to February 2020 at a large health system in the Los Angeles area revealed that a significantly higher number suffered respiratory complaints when compared to data from the previous six years.

These cases were not confirmed with tests, but the authors offered them as evidence of possible community spread in California during the previous winter.

· In July, a team reported their analysis of Canada's first 118 COVID-19 cases that had appeared between 25 January and 11 March 2020. All cases were related to travel, but only 7.6 per cent were connected to China. The rest – 92.4 per cent – had been to the United States, the Middle East, or Europe.

· On 9 September 2020, news came that an eighty-four-year-old man in Kent, England, who fell ill on 15 December 2019 and died on 30 January, had succumbed with the virus. His bereaved daughter angrily blamed China and its silence for his death.

· In October 2020, a team from Italy examined hundreds of blood samples, saved from a Milan study on cancer; more than 10 per cent tested positive for COVID-19 going back to October 2019 [*sic*]. At the request of WHO, the results were retested and confirmed in the Netherlands and another Italian lab. But the Dutch group quibbled over the degree of positivity. China received the results of retesting as positive and a sign of earlier spread; elsewhere, they are regarded as equivocal, meaning difficult to interpret.

· In December 2020, the American CDC reported a positive COVID-19 result from the throat of a four-year-old boy in Milan who had fallen ill on 21 November 2019 and had been swabbed on 5 December 2019 for suspected measles.

· In January 2021, wastewater samples placed COVID-19 in northern Italy by December 2019.

As these sporadic reports accumulated, the earliest date for the virus kept rolling back in time and its geographic range kept expanding. Meanwhile, the notion that the virus had escaped from the Wuhan laboratory, proved impossible to ignore, spurred on by some American leaders and the infodemic. On 30 October 2020, an in-depth study, led by the WHO, was launched with a virtual meeting. The team would locate and interview any still-living early sufferers to determine where and how they might have contracted the infection. Animals would also be studied, but virologist, Angela Rasmussen, cited in the top science journal, *Nature*, opined that it might be impossible to find the "smoking bat."

Politics entered this matter once again through the spin applied to each new observation. Eager to deflect blame for the origin of the virus and for delay in announcing its existence, China clung to the sporadic reports coming from other continents. Denying his own responsibility for the mismanagement and carnage at home, Donald Trump never relinquished his contention that it was a "Chinese virus," congratulating himself for banning flights from China and later Europe. Nor did he let up his attacks on the WHO and the notion that the organization was a "puppet of China." These postures further destabilized the already rocky relations in world trade and security.

Chinese authorities repeatedly postponed the WHO investigation; two scientists were denied entry to the country in early January 2021 after one had already set out. Finally, on 11 January 2021, amidst a new outbreak near Beijing, health authorities agreed to allow the seventeen-member WHO team to come to China. They planned to start work in the region of Wuhan. Arriving on 14 January, they began a fourteen-day quarantine, during which they held many virtual meetings with their seventeen Chinese counterparts. They requested and were given data of Chinese investigations into the live animal markets and early human cases. In the market, hundreds of swabs had been taken from organic and inorganic materials, sewers, and air vents. When their quarantine ended, the WHO team toured markets and met with officials and scientists; however, they were not provided with actual medical records of early cases and were forbidden to meet survivors of what might have been early outbreaks.

In advance, Chinese scientists had scoured records of more than seventy thousand hospitalizations in Wuhan during October 2019. They identified ninety-two still-living people who had suffered earlier respiratory illnesses with clinical courses like COVID-19. Two-thirds of those agreed to be tested for evidence of antibodies. The WHO team was simply told that all tests had been negative and that the remaining third had since died or refused testing. The team was frustrated by being unable to meet these patients or see data. China cited strict privacy laws for health records and the refusal of some individuals.

The final WHO report, released on 30 March 2021, immediately faced disappointment and criticism. The team admitted that they had not found the origin of the virus, nor could they specify the animal source, but they favoured a bat origin with a pangolin intermediary. They also dismissed as "extremely unlikely" the May 2020 claim of US Secretary of State Mike Pompeo, who cited "enormous evidence" (never revealed) that the SARS-COV-2 was a weaponized escape from the Wuhan laboratory. Readers of the late February Atlantic Council report, *Weaponized* (chapter 10), suspected that the WHO authors had been persuaded by one-sided rumours that caused them to include two hypotheses favoured by China: first, that the virus emerged elsewhere, perhaps in southeast Asia; and second, that it had entered markets via frozen food, probably fish. Suggestive evidence came from the genetic similarities of SARS-COV-2 to coronaviruses of bats in Thailand. Further evidence of early circulation outside China had been mounting, as shown above, and Wuhan had hosted several large international conferences in late 2019.

Notwithstanding the WHO report, the idea that the virus had escaped from the Wuhan lab persisted and had considerable currency among Americans. Various scientists continued to give it credence in the months that followed, hinting that the economic rival had been trying to weaponize the contagious germ in gain-of-function experiments, aiming to make it more contagious. In late May 2021, US intelligence revealed that three lab workers in Wuhan had fallen ill in November 2019; however, Wuhan had suffered an especially severe influenza season that year – one reason why it had been difficult to pluck out distinctive cases of the new disease. By mid-July 2021, even the WHO director general insisted that it was too early to rule out a lab accident, and he urged China to be more transparent.

The lab-leak theory escalated with a political twist. On 20 July 2021, the Republican senator Rand Paul accused NIH director Anthony Fauci of lying to Congress and killing four million people worldwide, because, he contended, NIH-funded scientists had collaborated with counterparts in Wuhan for gain-of-function research. Fauci retorted, "If there is any lying here, senator, it is you!" Perversely, in an academic climate that urges more international cooperation, it now seemed that any form of professional or scientific relationship with China had become an object of suspicion, if not a crime.

Many other theories continued to swirl around the virus and its origins. Against the lab-leak theory was growing recognition that it was nurtured by political hostility and ethnic targeting and the mounting evidence of SARS-COV-2 beyond China before the pandemic. Ethicists began to question the role of xenophobic orientalism in persistent efforts to situate the origins of COVID-19 in exotic Asian markets and laboratories. Linguists perceived denigration and racism in the persisting semantic associations with China.

Hoping to bring order and cooperation to the chaos, the WHO announced the formation of yet another advisory group in mid-October 2021. The new Scientific Advisory Group for the Origins of Novel Pathogens (SAGO) would re-examine the origin question and make recommendations for the future. It was to have representation from twenty-six nations, including China and the United States, as well as six members from the original fact-finding team. Its mandate included the need to address gaps in understanding of SARS-COV-2 and other emerging pathogens. The closed meetings began in late November 2021. Soon, the *omicron* wave washed over the late 2021 investigations, dragging the infodemic-laden scapegoating and protest along with it.

Fiction and fear became active participants in the ongoing search for the origins and control of COVID-19. Unfortunately, they will help to shape its future.

Looking Backward, Hoping Forward

I am wrapping up this book in May 2022, just thirty months since Canada's first case was reported in Toronto. Despite the fair-weather reopenings, COVID-19 is not over. It will be with us for a long time. Current estimates suggest that 70 per cent of people in England have had the infection--vaccinated or not; 43 per cent of Americans; 40 per cent of Canadians. Dr Theresa Tam is attempting to measure how many Canadians are long haulers, knowing that the syndrome may affect up to a quarter of those who suffered symptoms. It may also include hepatitis in children who had been infected with no symptoms at all.

Recent surveys suggest that 65 per cent of people in Africa have been infected while only 4 per cent are vaccinated; scientists are struggling to understand why. China, with its low incidence, is still attempting to maintain its zero-tolerance policy of massive but increasingly unaccepted controls in Shanghai and Beijing; models suggest *omicron* might still sweep the country. Russia's brutal war in Ukraine is raging; it might well be another epidemiological turning point in the pandemic, or it may not. All I know today is that the future is unknown and will have surprises. History can make us expect patterns or have hunches, but its knowledge is limited to the past. I happily leave the next months and years of the pandemic for others to record.

If we are lucky, COVID-19 will recede into an endemic problem with manageable parameters. In developed nations, the economy is already roaring back with unprecedented growth and rampant inflation, while shortages of every kind – workers and material – have emerged: automakers need chips; exhausted nurses are leaving the profession; container ships idle at sea waiting to unload; drug prices soar; the stock market is rocky. Economists suggest that it may be two years or longer before stability returns. Developing countries are likely to see more suffering and carnage before they make significant progress with vaccination.

In the heat of the moment, when those eager journalists contact historians of medicine, they ask for lessons from the past to help confront our present. Whether or not they realize it, those lessons are already embedded in our surroundings and institutions.

Every public-health organization, every sanitary measure, and all the recommendations that we are urged to obey, emerged as lessons from close encounters of the infectious kind in times past. They range from ancient provisions for shunning the infected, quarantining, and wearing protective garb. They include burning, burying, or destroying the remains and chattels of the dead. They incorporate the tenets of antisepsis and germ theory, the benefits of new medications, and the social, political, legal, and economic concepts of collective intervention on individual lives for a greater good.

But – just as we observed in the introduction – nature is endlessly inventive and it will always find new ways to get around the best barriers that we have imagined and created. At some point, these lessons from the past will be insufficient once again.

In the summer of 2020, a physician and philosopher from Iran, Pegah Mosleh, assembled an international team of scholars to reflect on political and philosophical questions raised by what he calls the "Corona Phenomenon," hoping to compile a book. He invited me to examine differences between fourteenth-century plague and COVID-19. At first, I was reluctant to accept the task; the differences seemed so obvious and so great. What use could it be? But in the end, I agreed, and the thought experiment was salutary, for me at least, if no one else. I realized that the comparison also worked well with the last great pandemic of influenza a century ago.

The first message is that epidemics (or pandemics) always follow a series of stages, a dramaturgic form, a notion described by historian Charles E. Rosenberg in his famous 1989 essay published in *Daedalus* during the context of AIDS. A new infectious disease begins with fear, panic, blaming, and scapegoating, often driven by xenophobia. It moves on to acceptance and hard work of management. It closes with resolution, new knowledge, and, alas, eventual forgetting. The new knowledge is what shapes the public-health barriers of the after time – barriers that will break with the next new challenge, and the cycle begins again.

COVID-19 has already displayed those stages, and like its predecessor pandemics, it has brought out the best and the worst in people. At the time of writing, we are somewhere in transition between the middle and final stages of this pandemic, having amassed mountains of information about the disease, its viral and social causes, and ourselves. Its tedious length serves as a reminder that pandemics past were not short, sharp episodes, as they tend to seem in retrospect. They, too, were protracted, sprawling, and multidimensional in time and space. Like plague, influenza, and AIDS, COVID-19 will continue – and like them, it will leave lasting changes.

The second message is that despite the similarities, many differences separate us from those distant and not-so-distant pandemics of plague and influenza. The differences make us lucky, relatively speaking.

For example, if we fall ill now, we have a much better chance of surviving than did our predecessors. It might be cold comfort for some; however, I marvel at the changes that came out of the previous century. We have concepts of hygiene to allow for care without contamination. We also have drugs to kill bacteria and viruses; oxygen stored in tanks; machines for artificial respiration; tests to identify carriers who have no symptoms; and brilliant biotechnology allowing us to define the invader chemically and make chemical vaccine and drug weapons against it. We have the internet and ever-improving communication platforms that not only allow the tracking of the infection in real time, but also keep people working, teaching, learning, meeting, and making music and art. Beyond the assurances provided by science and technology, many nations also have the concept and reality of social safety nets – the welfare state – the idea that relief packages are the responsibility of the collective. In this crisis, these great differences from pandemics of the past are things to celebrate.

As I write, COVID-19 has disrupted our lives, stolen employment, distorted our economy, and damaged our children's education. It has infected more than five hundred million and killed more than six million of the world's almost eight billion people. Furthermore, using the excess death method of counting (chapter 5), the WHO placed the number of dead in the first two years is closer to fifteen million; they included diagnosed and undiagnosed cases of COVID-19 and those who succumbed because of the pandemic's damage to health care and quality of life. These numbers are sobering. But they pale in contrast to the estimated twenty-five million who succumbed to plague in the fourteenth century when the globe's total population was about four hundred million – or to the five hundred million out of two billion who died of influenza in 1918, or the forty million who have died of AIDS so far. Those figures are mere estimates because, unlike us, no one could accurately measure the harm. Our abundant statistics will help shape the next defences and the next acute response. Stuck in this human predicament, yes, we have woe. But we also have more help than in the past, and we have hope; both are lessons – gifts – from the suffering and ingenuity of our predecessors.

Journalists also ask historians about the future. Being experts on the past and interpreters of the present, wise historians know not to predict the future. In fact, the late historian, John Burnham, once delivered a witty address, entitled "The Past of the Future of Medicine," in which he aired many pompous but entirely incorrect opinions foretold by hapless historians of medicine.

COVID-19 exposed enormous social, economic, and environmental problems around the world. The wildly varying and inconsistent responses, even within individual nations, proclaimed problems with federalism, where national standards and policies are complicated by philosophical notions of freedom, surveillance, and autonomy. We knew that these problems existed, but they were largely ignored. The pandemic drew attention to them because they made the experience far worse, aggravating numbers of cases, mortality, economic impact, and political strife.

I cannot predict the future, but I can hope. Indeed, I hope that our lives and interactions will revert to the old normal, of hugs, kisses, shaking hands, getting together in large crowds to enjoy conferences, sports, music, theatre, and gathering more intimately to celebrate the rituals of life – marriages, births, funerals, graduations, and dining en famille. These simple things, which we once took for granted and have missed so much, are returns – not changes. Even as we revert to in-person working, meeting, and playing, I hope that we will retain and rely often on the improved communication technologies for more egalitarian participation from a distance.

Nevertheless, based on what COVID-19 has forced us to observe in our responses to this challenge, I hope that several things will change – not only for confronting the next infectious debacle, but for improving our world in all the days ahead.

First, of course, we need to know that another pandemic will come at some point, sooner or later, and that we must never, ever, let down our guard. We need to bring back and sustain Canada's once admired system for vigilance and early alerts, Global Public Health Intelligence Network (GPHIN). We must ensure adequate, up-to-date stockpiles of PPE and drugs, and hone our understanding of sources and supply chains for the basic protective commodities that were so sadly lacking when this pandemic first hit. Every country needs an essential medicines list, reliable non-expired pharmaceutical resources, and clear direction for how to manage each drug shortage in a crisis. We need to remember that it was those ancient hygienic measures – not the drugs, not the vaccines – that stopped each prior outbreak in its tracks and that toppled each wave of COVID-19.

Second, beyond the spectacular scientific and technological achievements, we must study our experience and then remember how to roll out so many vaccinations at once. In some ways, we have a lot to learn from some African nations where mass immunization is handled well – but only when it is provided with vaccines. We need to fix the shameful, unequal distribution patterns of all immunization products.

Third, we need to closely examine the social traditions and economic policies that allowed us to neglect seniors and the disabled. Standards must be raised, homes renovated, and inspections restored. More acute-care beds could be useful. But health care is not only about hospitals, nurses, and doctors. Home care must be properly funded within the health-care system, not as a private option. Many people would choose to stay in their own homes where they would be happier, healthier, and ever so much safer – if only basic supports were available. The dignity of this choice should not be reserved for the wealthy.

Fourth, along the same lines, we need to recognize the inordinate burden that the pandemic placed on women, especially the mothers of young families. An enormous source of anxiety and financial loss came from the colossal effort of helping children follow online learning, while also caring for little ones and holding down outside jobs or working remotely. It is said that COVID-19 has set back the social and career advances of women by as much as two decades. National economies may have recovered quickly following prior pandemics, but there is scarring for individuals and families who many never recover at all. Programs may emerge to address this problem, but at the most basic level, it is obvious that publicly funded, accessible, and trustworthy childcare ought to be a human right – not only for parents and children, but also for the efficient operation of our entire economy. By the same token, teachers, and daycare providers – so long disrespected in terms of wages, working conditions, and benefits – must be recognized for the vital contribution that they, like parents, make to our youth and the functioning of our society and economy.

Fifth, COVID-19 exposed flagrant inequities. We must wake up to the idea that essential workers go well beyond the educated professionals in health-care facilities. They include the garbage collectors, grocery-store clerks, cleaners, personal support workers, meat packers, truck drivers, and transit employees, all of whom bravely risked their lives every day to keep our society afloat, often delivering services on wages so low that they needed more than one job. We need to recognize the role of racism and colonialism at the origin of the injustices. Similarly, we need to provide decent wages, sick pay, and furlough pay to ensure that no one with symptoms is tempted to leave home for work, possibly spreading infection, for lack of basic support.

Sixth, between nations and among scientists, we need to trust and talk to each other, collaborating and sharing information at every level of knowledge, policy, and management, something emphasized in the WHO report of February 2020 – something that is still not happening for COVID-19. Glimmers of novel possibilities in health care and the business practices of pharma production, research, clinical trials, and innovation are visible; we must not let them slide back. Open access and transparency in publication of news reports and research is another good step that, I hope, will endure. The ugly, selfish scramble over drugs and vaccines, pitting entitlement against poverty, exposes an even greater vulnerability. As a resource to turn this understanding into action, the WHO deserves respect and active participation. We are none of us safe, until we are all safe.

Seventh, through the work of academic researchers, the WHO, and other organizations, we must analyze the relative successes and failures in managing the pandemic – cases, deaths, and case fatality rates, as well as their relative economic impacts, to recognize emerging patterns of benefit in terms of political action and public policy. What can democracies learn from authoritarian states? What can countries without universal health care learn from those who have it? To what extent did the generous relief packages contribute to survival and recovery? How much and how often did any given region gain from willing contributions of volunteers? Who has been scarred?

Eight – and I'll stop at eight, though I can think of more – the clearing of the skies over China, Los Angeles, and New Delhi with the drop in air and vehicular traffic – temporary though it was – spelled out a stark lesson for us as stewards of the planet: human action on climate change can work and work quickly. COVID-19 presented numerous natural experiments for scientists to explore interactions between pollutants. The resultant calculations point to what changes are needed to reverse the trends. In February 2021, Dr Mike Ryan of the WHO said it well: "We are pushing nature to its limit. My fear is that ... someday ... our children will wake up in a world where there is a pandemic that ... could bring our civilisation to its very knees. We need a world that is more sustainable." Whether or not Ryan has ever heard of Grmek's concept of pathocenosis, caring for our planet is also health care. Ecologists have shown how maintaining biodiversity can prevent diseases, thereby connecting the dots between environmental degradation, climate change, and human health.

As much as we must prepare for another pandemic of an infectious disease, the next global scourge is already upon us, and we need to take steps to address it ... now.

For sources and suggested readings go to
https://www.mqup.ca/filebin/pdf/Sources_to_COVID-19.pdf

Index

Bhutan, 83, 125, 131, 183
Bible, 9, 28, 56, 180
Biden, Joe, 51, 87, 118, 153, 166, 171, 174
biodiversity, 65, 207
Biolyse, 118
BioNTech, 114. *See also*
Pfizer-BioNTech
biotechnology, 13, 113, 117, 162, 203
Black Lives Matter, 28, 63, 149
blame. *See* scapegoating
blood clotting, 96–7, 116, 175
BlueDot, 72–3
Bogach, Dr Isaac, 174
Bolivia, 112, 118
Bolsonaro, Jair, 39, 128, 142
booster shots, 181, 183, 184, 186, 187
border closures, 30, 48, 89, 141, 145,
150, 161, 178, 179, 194
Boston, 196
Boyle, Francis, 160
Bragg, W.H. and W.L., 58
Brazil, 46, 61, 88, 112, 124, 128, 173;
earliest cases in, 22, 39, 40, 196
breakthrough infection, 60, 173, 180,
181, 186, 190
Brexit, 50, 141, 169
British Columbia, 17, 30, 32, 134, 150,
157, 180, 182, 194
Burnham, John, 204

California, 19, 26, 129, 132, 150, 173,
196, 197
Campbell, William C., 97
Canada, 46, 63, 72, 79, 83, 88, 98, 100,
124, 134, 151, 161, 179; earliest cases
in, 17, 25, 30–8, 49, 50, 51, 197;
financial relief in, 38, 154–6, 188;
and infodemic, 160; later waves in,
133–8; *omicron* wave in, 185, 187, 190,
201; and vaccines, 109–10, 114,
117–18, 170–2, 176
Canada Worker Lockdown Benefit,
188. *See also* sick pay

Canadian Access to Medicines
Regime, 118
Canadian Armed Forces, 34, 146
Canadian Covid Care Alliance
(CCCA), 98, 99
Canadian Emergency Response
Benefit (CERB), 37
Canadian Institute for Health
Information (CIHI), 35, 63–4
Canadian Medical Association Journal,
34, 58
Canadian Recovery Sickness Benefit
(CRSB), 38
Canadian Trucking Alliance, 194
CanSino vaccine, 110, 116, 117, 170
Caribbean, 39, 150
case-fatality rate, 73, 124–5. *See also*
mortality
Castex, Jean, 176
CARES Act (*Coronavirus Aid, Relief,
and Economic Security Act*), 27–8,
89–90, 111, 153
CBC (Canadian Broadcasting
Company), 4, 6, 76, 143
cell-phone apps, 82
Centers for Disease Control (CDC), 17,
70, 72, 98, 197
Cercone, Philip, 6
chickenpox, 57, 109
children, 18, 37, 105, 107, 167, 187, 203,
205, 207; impact on, 145; symptoms
in, 123; and tests, 70, 71, 190; and
vaccines, 108, 109, 117, 162–3, 181,
182, 183, 184, 187. *See also* education
Chile, 40, 124, 142
China, 65, 67, 68, 69, 72, 74, 78, 83, 90,
100, 103, 125, 130, 151, 159, 160, 161,
201; first wave in, 11–24, 30, 46,
47–8, 49, 51, 195–200; map of, 16;
mortality in, 140, and vaccines, 112,
117, 165, 170, 174. *See also* Olympic
Games; zero tolerance
Chisholm, Brock, 19

Index